Succeed in College!

Walter Pauk
Cornell University, Emeritus

Ross J. Q. Owens

Houghton Mifflin Company
Boston New York

Vice President and Publisher: Charles Hartford
Executive Editor: Richard Stratton
Senior Development Editor: Rita Lombard
Editorial Associate: Liz Hogan
Production Editor: Claudine Bellenton
Senior Manufacturing Coordinator: Marie Barnes
Senior Marketing Manager: Katherine Greig
Marketing Coordinator: Alexandra Shaw

Cover image: © Galen Rowell/Mountain Light
Cover description: "Self-portrait at dawn, Torres del Paine, Patagonia, Chile"

Printed in the U.S.A.

ISBN: 0-618-47237-1

1 2 3 4 5 6 7 8 9-VH-09 08 07 06 05

To the Student...

This book was designed to help you achieve success, not only in your introductory physical science course, but also throughout your college career. The information in this book was derived from the best-selling college study skills text, *How to Study in College*, Eighth Edition, by Professor Walter Pauk of Cornell University. The chapters present effective methods for mastering areas of interest to every student—managing time, taking good notes, learning from textbooks, and performing well on tests. Millions of students have used Walter Pauk's text in its first seven editions. Now you can benefit from the Pauk approach to study skills.

Succeed in College! Has been prepared exclusively to accompany *An Introduction to Physical Science*, by Shipman, Wilson, and Todd, and is available only with this textbook. Here's to your success in college physical science—and beyond!

Contents

PART I

Permanent Skills

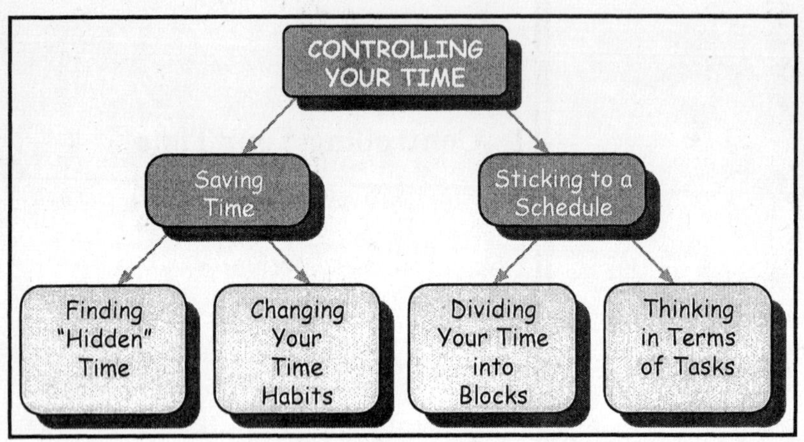
CONTROLLING YOUR TIME
Saving Time
Sticking to a Schedule
Finding "Hidden" Time
Changing Your Time Habits
Dividing Your Time into Blocks
Thinking in Terms of Tasks

CHAPTER 1

Controlling Your Time

Perhaps the most valuable result of all education is the ability to make yourself do the thing you have to do, when it ought to be done, whether you like it or not.

Thomas Huxley
(1825–1895),
English biologist

Time flies, but that's no reason for you to go through each day simply "winging it." Through conscientious use of time and commonsense planning, you can make the most of your day. This chapter ticks off the important elements of time management, including:

- **Finding "Hidden" Time**
- **Changing Your Time Habits**
- **Dividing Your Time into Blocks**
- **Thinking in Terms of Tasks**

How important is time management to your academic success?

Time is a precious and irreplaceable commodity. As she lay on her deathbed, Queen Elizabeth I of England (1533–1603) reportedly said, "All my possessions for a moment of time." How you use time can determine your success or failure in college. If you use time wisely, you'll prosper. If you use it poorly, you'll fail in the job you came to do. That's why the management of time is the number-one skill to master in college.

What did the Fordham study find about how students use their time?

Yet students frequently squander time. A survey conducted at Fordham University found that college freshmen spent roughly one-third of their waking hours engaged in social activities or idle leisure during a typical weekday. This "free time" amounted to nearly twice the time they spent studying. And on the weekend the ratio of social and idle leisure time to study time for the same group was almost six to one!

How can you gain extra time?

Although the students in the survey seemed to waste time routinely, you needn't put yourself in the same position. You can gain extra time in two ways: (1) by doing a job in less time than usual, and (2) by taking control of your time. The first way requires you to study more efficiently, and this book provides a great many techniques to help you do just that. The second way requires you to save time by reclaiming "hidden" time and improving your time habits and to better manage your time by sticking to a schedule.

Saving Time

How can you put your time to better use?

All of us have claimed that we don't have enough time to accomplish what we need to do. But the fact is that everyone is allotted the same amount of time: twenty-four hours a day. Many of us allow a lot of this time to go to waste by failing to realize it is available in the first place. In addition, it's often our day-to-day habits, activities we no longer notice, that save time or waste it. You can put your time to better use by pinpointing areas of "hidden" time and cultivating time-saving habits.

Finding "Hidden" Time

What is "hidden" time?

There's a lot of valuable time in your day that is being overlooked, simply because you didn't realize it was time you could use. For those who flush tiny slivers of soap down the drain or toss small scraps of cloth into the wastebasket, there are others who combine those slippery bits and pieces into a whole new bar or stitch discarded shreds into a comfortable quilt. Think of all the time you spend standing in line or even waiting for a traffic light to change. If you could find ways to make use of this "hidden" time, you could almost certainly add hours to each week.

Carry Pocket Work

How does pocket work use hidden time?

Many situations may leave you with a few moments of unexpected free time—a long line at the bank or supermarket, a delayed bus or train, a wait at the doctor's office, a lunch date who arrives late. If you make a point to bring along a book, a photocopied article, index cards on which you've written key concepts, vocabulary words, or formulas, you'll be able to take advantage of otherwise frustrating experiences.

Use Your Mind When It's Free

When is your mind free for studying?

Some activities may afford an overlooked opportunity for studying if you're prepared. For example, if you're shaving, combing your hair, or washing dishes, there's no reason you can't be studying at the same time. Since many of us tend to "zone out" in such situations, they are excellent opportunities to use time that might otherwise be squandered. Attach small metal or plastic clips near mirrors and on walls at eye level. Place a note card in each clip. Do a problem or two in math or master some new vocabulary words as you eat a sandwich at work.

Record Study Information

How can recording information use hidden time?

Another way of using hidden time is by listening to information you've recorded on audiocassettes or MP3 files or burned onto CDs. Recorded information enables you to keep studying in situations where you're moving about or your eyes are otherwise occupied, such as when you're getting dressed or driving. In addition, recorded information can provide a refreshing change from written material.

Employ Spare-Time Thinking

What is spare-time thinking?

You can make the most of the moments immediately before or after class by recalling the main points from the last lecture as you're heading to class or by quickly recalling the points of a just-completed lecture as you're leaving class.

Use Your Subconscious

How can you use your subconscious to save time?

At one point or another, you have awakened during the night with a bright idea or a solution to a problem that you had been thinking about before bedtime. Your subconscious works while your conscious mind is resting. If you want to capture the ideas or solutions produced by your subconscious, write them down as soon as you wake up; otherwise, they'll be lost. Many creative people know this and keep a pad and pencil near their beds. For example, Nobel Prize winner Albert Szent-Györgyi said, "I go to sleep thinking about my problems all the time, and my brain must continue to think about them when I sleep because I wake up, sometimes in the middle of the night, with answers to questions that have been eluding me all day."[1]

Changing Your Time Habits

What's a good way to begin saving time?

Habits, by their very nature, are things we do routinely without even thinking. Most of us are unaware of our habits unless someone or something draws attention to them. A good way to take inventory of your time habits is by keeping a daily activities log. From the time you wake up to the time you go to sleep, note all your major activities, the time you started and finished each, and the time each activity consumed. With your day itemized on paper, you can gain a clearer picture of where your time is being spent and where it's being wasted. The activities log in Figure 1.1 shows one student's daily routine and how he decided to put his time to better use.

[1] Originally published in *Some Watch While Some Must Sleep,* by William C. Dement, as a volume in The Portable Stanford series published by the Stanford Alumni Association. Copyright © 1972. Reprinted by permission of the Stanford Alumni Association.

Time Start	Time End	Time Used	Activity - Description
7:45	8:15	:30	Dress
8:15	8:40	:25	Breakfast
8:40	9:00	:20	Nothing
9:00	10:00	1:00	Psychology - Lecture
10:00	10:40	:40	Coffee - Talking
10:40	11:00	:20	Nothing
11:00	12:00	1:00	Economics - Lecture
12:00	12:45	:45	Lunch
12:45	2:00	1:15	Reading - Magazine
2:00	4:00	2:00	Biology Lab
4:00	8:00	4:00	Work (includes 1/2 hour dinner break)
8:00	8:50	:50	Study - Statistics
8:50	9:20	:30	Break
9:20	10:00	:40	Study - Statistics
10:00	10:50	:50	Chat with Bob
10:50	11:30	:40	Study - Accounting
11:30	11:45	:15	Ready for bed
11:45	7:45	8:00	Sleep

Paste index cards on mirror; laws of economics; psychological terms; statistical formulas. Study while brushing teeth, etc.

Look over textbook assignment and previous lecture notes to establish continuity for today's psychology lecture.

Break too long and too soon after breakfast. Work on psychology notes just taken; also look over economics assignment.

Rework the lecture notes on economics while still fresh in mind. Also, look over biology assignment to recall the objectives of the coming lab.

Break is too long.

Good as a reward if basic work is done.

Insufficient time allotted, but better than no time.

While brushing teeth, study the index cards. Replace cards that have been mastered with new ones.

Figure 1.1
Record of One Day's Activities and Suggestions for Making Better Use of Time

How can you use what you learn from your daily log?

Once you have the concrete evidence of a daily activities log before you, you can begin to eliminate your time-wasting habits and develop or reinforce the time-saving ones.

Defy Parkinson's Law

What can you learn from Parkinson's Law?

Parkinson's Law says that work expands to fit the time allotted.[2] To avoid running out of time, work Parkinson's Law in reverse: For each task, set a deadline that will be difficult to meet, and strive to meet that deadline. Each time you achieve your goal, reward yourself with some small but pleasant activity. Take a break. Chat with a friend. Stroll around the room. Have a special snack, such as a bag of peanuts (keep it in your desk, to be opened only as a reward). If you fail to meet a deadline, don't punish yourself. Just hold back your reward and set another goal. Positive reinforcement is powerful in effecting a change in behavior.

---

[2] C. Parkinson, *Parkinson, the Law* (Boston: Houghton Mifflin, 1980).

Obey Your Alarm Clock

Why is it so important to obey your alarm?

How many times do you hit the snooze button on your alarm clock before you finally get out of bed? Even one time is too many. Set your alarm for the time you want to get up, not for the time you want to start getting up. If you can't obey your alarm, you'll have a hard time sticking to your time schedule. After all, it doesn't even buzz.

Limit Your E-mail and Internet Time

What can you do to limit your e-mail and Internet time?

As marvelous as they both can be, e-mail in particular and the Internet in general can be tremendous "time sinks," swallowing up hours in a typical day. Rather than checking it constantly, designate specific times during the day when you read and send e-mail. It's true that e-mail has sped up communication, but it's a rare message that can't wait a few hours before being read or sent. The same applies to any Web surfing you may do, whether for schoolwork or pleasure. Time has a tendency to fly by as you click from one link to the next. You can help keep things under control by setting a timer when you surf and returning to your studies when the timer goes off.

Take "Time Out"

What is the value of taking time out?

Reward yourself with regular short breaks as you work. Learning in several small sessions, rather than in one continuous stretch, increases comprehension. In one study, students who practiced French vocabulary in three discrete sessions did 35 percent better on an exam than those who tried to learn the words in one sitting.[3] So take a breather for ten minutes every hour, or rest for five minutes every half-hour. Whichever method you choose, keep your breaks consistent. This way, you'll study with more energy and look forward to your regular rests. When you return to your desk, you'll find that you feel more refreshed.

Listen to Your Body

How can understanding circadian rhythms affect your scheduling?

All of us are subject to circadian rhythms. That is, we have periods when we're most wide awake and alert and other periods when we're sluggish or drowsy. In

---

[3] Kristine C. Bloom et al., "Effects of Massed and Distributed Practice on the Learning and Retention of Second-Language Vocabulary," *Journal of Educational Research* 74, no. 4 (March–April 1981): 245–248.

general, we're sleepiest a few hours before dawn and twelve hours later, in mid-afternoon. In keeping with these natural cycles, we're widest awake about every twelve hours, usually at mid-morning and again in mid-evening. Knowing this can help you plan the day's activities more strategically.

1. *Schedule cerebral tasks for mornings and evenings.* Reading, writing, problem solving, and other "thinking tasks" should be done when you're likely to be most alert.
2. *Save active behavior for mid-afternoon.* Fieldwork, lab work, exercise, and errand running are best done at this time of day, when more sedentary activities may make you feel drowsy. If you're not a heavy coffee drinker, a cup of coffee might get you through the afternoon slump.
3. *Resist the temptation to sleep in on the weekends.* Changing your sleep schedule on the weekends can have a chain reaction effect on the following week. You may find yourself feeling jet-lagged on Monday or Tuesday if you sleep in on Saturday or Sunday.
4. *Read in the morning; review in the afternoon.* Scientists have discovered that short-term memory peaks at about nine o'clock in the morning and that long-term memory is strongest at about three o'clock in the afternoon.

Keep a Notepad Handy

How can a notepad improve your time habits?

As we'll see in the next chapter, stray thoughts and worries can really cut into your ability to concentrate. They can also eat up valuable time. The best way to deal with these distractions is to get them out of your head and onto a piece of paper as quickly as possible, then return to your work. Writing these thoughts down can result in an additional time saving bonus. Getting started is often the hardest part of shifting from one activity to another. Your jottings may provide a needed jump-start to overcome the inertia that seems to characterize the outset of a new activity. If so, they may save you some valuable time. Here is an example from the note pad of one student who, while working on a calculus assignment, came up with a topic for an upcoming paper. As soon as she finished her calculus, she was able to begin doing preliminary research on the topic without delay.

Call Prof. Singh about make-up test.
Check Campbell book for discussion of brain laterality.
What about "Earthquake Prediction" as possible paper topic?
Look up definitions for leftover vocabulary cards.
Tennis at 6 tonight, not 7!

Of course, although spontaneous scribblings can sometimes provide guidance as you're moving from one task to the next, a far better strategy is to come up with a schedule and stick to it.

Sticking to a Schedule

What is the function of a time schedule?

A time schedule is a game plan, a written strategy that spells out exactly what you hope to accomplish—during a day, a week, or even the entire term—and how you plan to do it. Committing yourself to planning and keeping to a schedule can seem a bit frightening at first, but following a schedule soon becomes a source of strength and a boon to your life. There are several benefits to a schedule.

How will a schedule provide greater control?

A schedule provides greater control. A thoughtfully constructed time schedule can increase your sense of control in four ways. First, because your schedule is written down, your plans seem more manageable. You can start working without delay. Second, you know you'll study all your subjects—even those you dislike—because you've allotted time for them in your schedule. There's less of a temptation to skip disliked subjects when study time has already been allotted for them. Third, a schedule discourages laziness. You've got a plan right in front of you, and that plan says, "Let's get down to business!" Fourth, you can schedule review sessions right from the start and avoid last-minute cramming for tests.

How does a schedule encourage relaxation?

A schedule encourages relaxation. At the same time, because your plan is written down instead of floating around in your head, your mind is freed for other things. There's no time wasted worrying about what to do next. It's all there on paper. There's no guilt either. Both work and play are written into your schedule. This means that when you take a break, you know you deserve it.

Why are some students reluctant to use time schedules?

Despite these benefits, many students are reluctant to start using time schedules. They feel not only that a schedule will do them little good but also that keeping track of time will turn them into nervous wrecks. Neither worry is warranted.

How does a schedule save you time?

A schedule saves time. Yes, it takes time to devise a schedule, but that time is rewarded. You will be able to shift smoothly from one activity to another, without wondering what to do next.

How does a schedule provide freedom?

A schedule provides freedom. Scheduling frees you from time's control. The people you see dashing from class to library to gym, or eating lunch on the run, are slaves to time. The students who schedule time, who decide how time will be used, are the masters of time.

In what way does scheduling increase flexibility?

A schedule increases flexibility. Disorganized people often waste so much time that there's no room for flexibility. People who use schedules free their time for a variety of activities and are therefore more flexible.

Which schedule type is best: time-based or task-based?

If you are a full-time student or have considerable control over the hours in your day, your best bet is to rely on traditional schedules that divide your time into manageable blocks. If, however, you are juggling your academic life with the responsibilities of a job or the demands of raising a family, you're probably better off using schedules that focus on tasks instead of on time. Each approach has advantages, but both provide an opportunity to tackle tasks with a genuine game plan instead of flailing at them haphazardly.

Dividing Your Time into Blocks

What does the story of the sticks illustrate about time?

A father once tied a bundle of small, thin sticks together with a strand of twine, handed the bundle to his youngest son, and said, "Son, break these sticks in half." The boy used his hands and knees but could not break the bundle. Sadly, he handed it back to his father. Without a word, the father untied the twine, and using only his fingers, snapped each stick one by one.

What is the advantage of dividing your time into blocks?

When the sum total of your obligations and academic assignments seems overwhelming, it helps immensely to split them up into small, manageable units. By dividing each day into blocks, time schedules break up your responsibilities and allow you to deal with them one by one. Assigning a block of time to each activity ensures that you will work at peak efficiency.

What are the components of the three-part scheduling system?

Using time blocks, you can create a three-part scheduling system consisting of (1) a master schedule, (2) a weekly schedule, and (3) a daily schedule. Each plays an integral role in managing your time. The master schedule serves as a basic structure for organizing your activities; the weekly schedule adds specific details to the master schedule; and the daily schedule puts the weekly schedule into a portable form. Although each schedule performs a different function, all three follow the same basic scheduling guidelines:

What is the best way to use big blocks of time?

1. *Don't waste big blocks.* If you have a big block of time, use it for a big assignment. There's a strong tendency to say, "I'm going to get these smaller assignments out of the way before I tackle the big assignment." This is a poor decision. Instead, use the large block of time for a large and time-intensive assignment, and save your small assignments for the little slivers of time.

When it comes to studying, what is prime time?

2. *Study during prime time.* For most of us, prime time is daytime. In fact, research has shown that each hour used for study during the day is equal to an hour and a half at night. Even so, you may find that you have dead hours during the day when you are less productive than you'd like to be. Schedule less-demanding tasks for these hours.

How does the type of class influence when you should study for it?

3. *Study before recitation classes and after lecture classes.* A study session before a recitation or discussion class (a foreign language course or a psychology seminar, for example) helps warm you up. When you walk into class, the material is fresh in your mind. For lecture classes, use the time immediately after class to fill in any gaps in your notes and to review the information you've just learned.

Why should you shy away from making your schedule too detailed?

4. *Don't let details tie your hands.* Account for all your time, but don't be overly detailed. The time you'd take to make an overly meticulous schedule can be better used in studying a subject directly, and the chances of your following such a strict plan are slim.

What sorts of nonacademic activities belong in your schedule?

5. *Include nonacademic activities.* Always set aside time for food, sleep, and recreation as well as the other activities of your life. Cheating yourself out of a meal, a good night's sleep, a swim, a family get-together, or a meeting with friends won't save you time in the long run. In fact, it may cost you time

because all these activities are necessary for your overall mental and physical wellness. Make your plan for living, not just for studying.

Lay a Foundation with a Master Schedule

What is the purpose of the master schedule?

A master schedule provides an agenda of fixed activities around which your varying activities are arranged. Unless changes occur in your basic program, you need to draw up a master schedule only once per term.

What does a master schedule look like?

A master schedule grid lists the days of the week at the top and the hours of the day down the left side. The boxes in the grid are filled in with all your required activities: sleep, meals, job, regular meetings, community activities, sports, and, of course, classes. The empty boxes that remain represent your free time. Figure 1.2 provides an example of a typical master schedule.

	Mon.	Tues.	Wed.	Thurs.	Fri.	Sat.	Sat.
7-8	←		Dress and Breakfast		→		
8-9	Bio-Sc		Bio-Sc		Bio-Sc	Dress & Breakfast	
9-10		P.E.		P.E.		P.E.	Dress & Breakfast
10-11	History		History		History		
11-12		Spanish		Spanish		Spanish	
12-1	←			Lunch			→
1-2	Math	Computer Lab.	Math	Computer Lab.	Math		
2-3	English		English		English		
3-4		Work study Prog.		Work study Prog.			
4-5	Work Study		Work Study		Work Study		
5-6							
6-7	←			Dinner			→
7-8							
8-9							
9-10							
10-11							
11-12	←			Sleep			→

Figure 1.2
A Master Schedule with Work

How does a master schedule help?

A master schedule, on a five-by-eight-inch card taped over your desk or carried in your notebook, unclutters your mind. More important, it enables you to visualize actual blocks of time into which you can fit necessary activities.

Account for Changing Details with a Weekly Schedule

What is the purpose of a weekly schedule?

The weekly schedule takes over where the master schedule leaves off. To construct it, photocopy or print out another copy of your master schedule, and fill in the empty blocks with the activities you have planned for the upcoming week. If you have a math test on Friday, for example, you will need to schedule a little extra study time for math. Next week you may be assigned a research paper. If so, you'll probably want to leave space in your schedule for library or Internet research. The weekly schedule helps you adapt your time to your changing priorities. Keep it posted by your desk or pasted on the inside cover of your notebook.

A sample weekly schedule is shown in Figure 1.3. The lists that follow show how the guidelines for scheduling were used to set it up.

Monday Through Friday/Saturday

7–8 A.M.	Avoid the frantic dash and the gobbled (or skipped) breakfast by getting up on time.
12–1 P.M.	Take a full, leisurely hour for lunch.
5–6	Relax before dinner—your reward for a day of conscientious work.
7–9	Keep up with current notes and assignments through systematic studying.
9–10	To forestall cramming at quiz and examination times, give some time every day to a review of previous assignments and ground covered to date.
10	A cease-study time of 10 P.M. provides an incentive for working hard during the day and early evening.
10–12	Devote some time every day to reading books that truly interest you. Recreational reading and conversation help you unwind for a good night's sleep.

Tuesday/Thursday/Saturday

8–9 A.M.	Because chemistry (10–11) is your hardest subject, build your morning study program around it. An hour's study before class will make the class period more meaningful.
11–12 P.M.	Another hour's study immediately after chemistry class will help you remember the work covered in class and move more readily to the next assignment.

Special

Tuesday	2–5 P.M.	Library: paper
Sunday	7–9 P.M.	English paper
		For some assignments you will need to schedule blocks of time to do research or to develop and follow up on ideas.
Saturday	From noon on, Saturday is left unscheduled—for recreation, for special projects to which you must devote a concentrated period of time, for extra work on difficult subjects, for thorough review.	
Sunday	This is your day until evening. Study history before you go to bed because it is the first class you'll have on Monday morning.	

	Mon.	Tues.	Wed.	Thurs.	Fri.	Sat.	Sat.
7-8	← Dress and Breakfast →						↑ Religious Service, Recreation, Conversation, Recreational Reading ↓
8-9	History	Study Chem.	History	Study Chem.	History	Study Chem.	
9-10	Study History	Phys. Ed.	Study History	Phys. Ed.	Study History	Phys. Ed.	
10-11	Study French	Chem.	Study French	Chem.	Study French	Chem.	
11-12	French	Study Chem.	French	Study Chem.	French	Study Chem.	
12-1	← Lunch →						
1-2	Math	Film-making	Math	Film-making	Math	↑ Recreation, Conversation, Special Projects, Reading, Extra Work on Difficult Subjects Through Review ↓	
2-3	Study Math	Library; Paper	Study Math	↑	Study Math		
3-4	Study English	Library; Paper	Study English	Chem. Lab.	Study English		
4-5	English	Library; Paper	English	↓	English		
5-6	← Recreation →						
6-7	← Dinner →						
7-8	Study English	Study Math	Study English	Study Math	Study English		English Paper
8-9	Study French	Study History	Study French	Study History	Study French		English Paper
9-10	Review English	Review French	Review History	Review Math	Review English		Study History
10-11	← Recreational Reading →						
11-12	← Conversation, Sleep →						

Figure 1.3
A Detailed Weekly Schedule Based on a Master Schedule

Provide a Portable Game Plan with a Daily Schedule

What goes into a daily schedule?

A daily schedule is a brief yet specific list of the day's tasks and the time blocks you plan to accomplish them in. You should be able to fit all this information on an index card that you can carry around with you all day. Make up your daily schedule each night before you go to bed. Once you have put your worries and concerns on paper, your mind will be free for sleep. You will also have thought through your day and will be better prepared when the morning comes. Figure 1.4 shows one student's daily schedule and explains why it is effective.

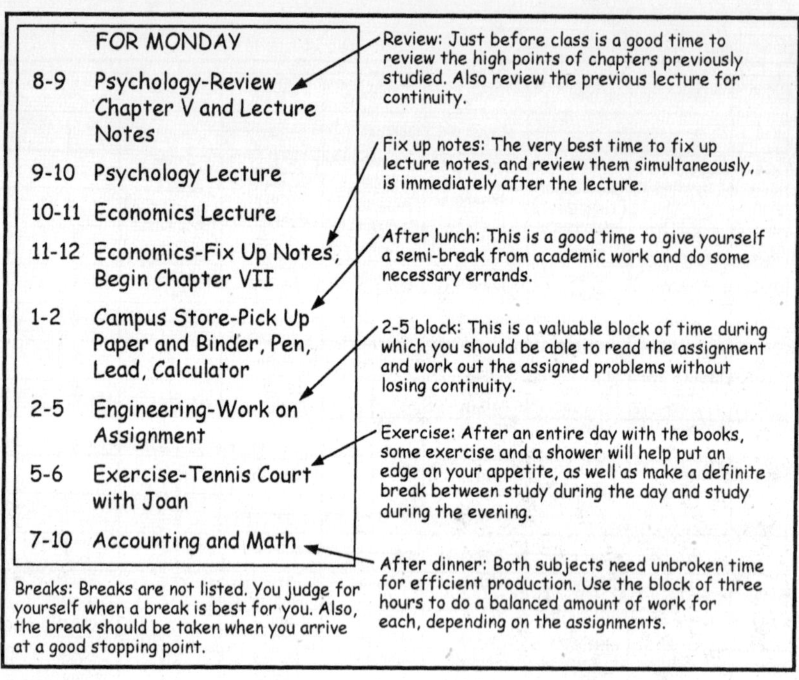

Figure 1.4
A Daily Schedule

Use Scheduling Tools If You Feel Comfortable with Them

What are the pros and cons of special scheduling tools?

These days, you can choose from a growing variety of tools and utilities, from computer software to personal digital assistants to Internet applications to hefty loose-leaf notebooks, all designed to make scheduling your time both easier and more intelligent. These tools may provide the breakthrough that some students need to finally begin to appreciate the importance of keeping a schedule. For other students, they are an elaborate, costly, time-intensive distraction that does little to control or organize time. Index cards and blank sheets of paper may not be all that sophisticated, but they are extremely inexpensive and breathtakingly simple, and they can usually provide all you need to get a firm grasp on your available time. If you swear by your scheduling software or your store-bought appointment book, that's fine. As long as you keep in mind that your goal is managing time, you can make just about any system work. The essential component is not the tool, but rather the person who is using it.

Thinking in Terms of Tasks

When is it preferable to use task-based scheduling?

Because of the way a schedule based on time blocks divides the day into manageable bite-sized segments, it is often preferable for most students. But some of us don't have the luxury of predictably structuring our days from top to bottom. If your days are largely unpredictable or your free time is fluid, you may need schedules that emphasize tasks instead of blocks of time. In addition, long-term assignments, which can't always be squeezed into time blocks, may require a task-based approach.

How did Ivy Lee demonstrate the value of tasks?

Anyone who's ever had the satisfaction of systematically crossing items off a lengthy to-do list has at least an inkling of the effectiveness of a task-based approach. During the first half of the twentieth century, legendary management consultant Ivy Lee demonstrated just how effective—and lucrative—it could be to view things in terms of tasks.

Charles Schwab, then chairman of the Bethlehem Steel Company, challenged Lee: "Show me a way to get more things done with my time, and I'll pay you any fee within reason." Lee thought for a while, then said,

- Every *evening* write down the six most important tasks for the next day in order of priority.
- Every *morning* start working on task #1 and continue until you finish it; then start on task #2, and so on. Do this until quitting time and don't be concerned if you have finished only one or two tasks.
- At the end of each day, tear up the list and start over.

When Schwab asked how much he owed for this advice, Ivy Lee told him to use the plan for a few weeks, then send in a check for whatever he thought it was worth.[4] Three weeks later, Lee received a check for $25,000, which is equal to about $250,000 in today's dollars!

By using good judgment, you can allot the bulk of your time to getting top-priority tasks done, yet not ignore other tasks with due dates.

What sort of schedules do you need for a task-based approach?

Of course, just as a daily schedule would provide only a limited view of your tasks, a single to-do list is not sufficient for a task-based approach to help you through the entire semester. Whether it hinges on time blocks or tasks, a three-part scheduling system still makes sense. If you're focusing on tasks, you'll need (1) a task-based master schedule and (2) a weekly assignment-oriented schedule in addition to (3) a daily to-do list.

Develop a Task-Based Master Schedule

How does a task-based master schedule differ from a standard master schedule?

A task-based master schedule enables you to keep track of one or more assignments or goals over an extended period of time. Figure 1.5 provides an example of a task-based schedule. Across the top of the schedule, instead of the days of the week, list the major goals you hope to accomplish or the assignments you plan to complete. Deadlines for subgoals may be written down the left side, where the hours of the day would normally be written in a standard master schedule.

How do you divide your goals into subgoals?

Now divide each goal or long-term assignment into manageable subgoals. List them in a column beneath the task they refer to. For example, if you've been assigned a research paper, you may arrive at the following subgoals: do preliminary research, choose topic, plan outline, conduct research, complete first draft, and revise first draft. As you reach each milestone on the way to completing your assignment, cross it off your schedule. As you do, you provide yourself with visual evidence of and positive feedback for the progress you've made.

---

4 T. W. Engstrom and R. A. Mackensie, *Managing Your Time* (Grand Rapids, MI: Zondervan, 1967).

	Psychology Research Paper April 21	Train for Amateur Triathlon May 1	Self-Paced Computer Course
Feb. 7	Select Three Topic Ideas	Up Minimum to 60 Laps	Complete Ch. 1-3
Feb. 10	Do Preliminary Research	Try Ride Up Satyr Hill	
Feb. 14	Make Final Topic Choice	Run 30 Miles Per Week	Complete Ch. 4-6
Feb. 18	Complete Bibliography		
March 15	Finish First Draft		Mid-Term Exam
March 18	Begin Rewriting		
April 21	Paper Due		Final

Figure 1.5
A Task-Based Master Schedule

Make Your Weekly Schedule Assignment-Oriented

When should you use an assignment-oriented weekly schedule?

If the span of your goal or assignment is a week or less, you can use an assignment-oriented weekly schedule as a supplement to your master schedule. Figure 1.6 shows such a schedule. The format is simple. Draw a horizontal line to divide an eight-and-one-half-by-eleven-inch sheet of paper in half. In the top half, list your subjects, assignments, estimated study times, and due dates. Then, with the due dates and estimated times as control factors, check your master schedule for available time. Allocate enough hours to do the job, and write them on the appropriate line on the bottom half of the sheet. Stick to your schedule. As long as you give study hours top priority, your remaining hours will be truly free.

Subject	Assignment	Estimated Time	Date Due	Time Due
Electronics	Chapter V-32 pp. - Read	2 hr.	Mon. 13th	8:00
English	Paper to Write	18 hr.	Mon. 20th	9:00
Math	Problems on pp. 110-111	3 hr.	Tues. 14th	10:00
Industrial Safety	Make Shop Layouts	8 hr.	Fri. 17th	11:00
Computer Graphics	Generate Slide Presentation (2-4 slides)	6 hr.	Fri. 17th	1:00
Electronics	Chapter VI - 40 pp. - Read	2 1/2 hr.	Wed. 22nd	8:00

Day	Assignment	Morning	Afternoon	Evening
Sun.	Electronics-Read Chap V English-Find a Topic			7:30-9:30 9:30-10:30
Mon.	English-Gather Notes Math-Problems		2:00-6:00	7:00-10:00
Tues.	English-Gather Notes Industrial Safety	8:00-10:00	3:00-6:00	7:00-10:00
Wed.	English-First Draft Computer Graphics		2:00-6:00	7:00-10:00
Thurs.	Industrial Safety English-Paper Computer Graphics	8:00-10:00	3:00-6:00	7:00-10:00
Fri.	English-Final Copy Electronics		2:00-6:00	7:00-9:30
Sat.				

Figure 1.6
A Weekly Schedule Based on Assignments

Turn Your Daily Schedule into a To-Do List

How should you adjust your daily schedule when your time is unpredictable?

Your daily study schedule should simply be a list of things to do arranged in order of priority on an index card. In this case, assigning specific times is likely to lead to frustration.

Figure 1.7 shows a typical daily list. To be successful, you need a sense of urgency about referring to your list and studying whenever an opportunity presents itself. Cross off the tasks as you complete them.

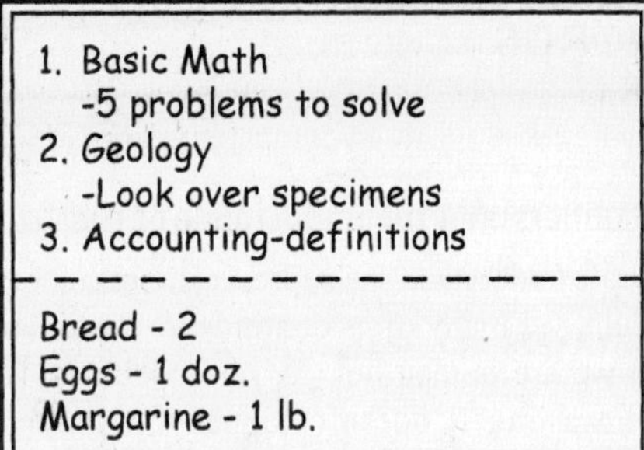

Figure 1.7
A Things-to-Do List

What is the Pareto Principle?

Use the *Pareto Principle* to help draw up your list. Named after Vilfredo Pareto (1848–1923), an Italian economist and sociologist, the Pareto Principle states that the truly important items in any group constitute only a small number of the total items in the group. This principle is also known as the *80/20 rule.*[5]

What are some examples of the 80/20 rule?

For example, in almost any sales force, 80 percent of the business is brought in by 20 percent of the salespeople. In any committee, 80 percent of the ideas come from 20 percent of the members. In a classroom, 80 percent of the teacher's time is taken up by 20 percent of the students.

In any list of things to do, 80 percent of the importance resides in 20 percent of the list. In a list of ten items, 80 percent of the list's value lies in two items, which constitute 20 percent of the list. Because of the Pareto Principle, always list the most important items first. Then, if you accomplish only the first few items, you will have accomplished the most important tasks.

Keep the Pareto Principle in mind whenever you make up a list or a schedule or must decide which subject to study first. Apply the principle by listing first things first.[6]

---

[5] Reprinted with the permission of Scribner, a Division of Simon & Schuster, from *Getting Things Done,* Revised and Updated Edition, by Edwin C. Bliss. Copyright © 1976, 1991 by Edwin C. Bliss.

[6] B. Eugene Griessman, *Time Tactics of Very Successful People* (New York: McGraw-Hill, 1994).

FINAL WORDS

Why do people who understand the importance of time continue to waste it?

Let's face it. Most people understand the supreme importance of time and the value of sticking to a schedule, yet many of us continue to waste time needlessly. Why? Perhaps it is because we are human beings, not machines. Learning to use time wisely after years of wasting it doesn't happen by simply flipping a switch. Controlling your time is ultimately a matter of developing the right mindset. If the dying words of Queen Elizabeth and the compelling evidence of this chapter haven't been enough to convince you to change your time-wasting ways, perhaps the following tidbits about time will finally make things click.

Ten Valuable Tidbits About Time

1. One of America's greatest composers, jazz giant Duke Ellington still wasn't ashamed to admit that "Without a deadline, I can't finish nothing'."
2. If you've got a long list of things to do but you're spinning your wheels, start with the easier tasks. Getting one done makes the next one easier. Build momentum.
3. Be a contrarian. Go to the library when almost nobody is there. Get into the dining-hall line before the crowd. Get the reserved books before the line forms. The time you save will add up quickly.
4. Make decisions wisely by asking, "What are the alternatives?" Make a list of the alternatives, and then put pluses and minuses alongside them. Learn this process. It will save lots of time.
5. Don't spread yourself thin by attempting to become an "information junkie." This scattershot approach takes up a great deal of time and can still leave you feeling stressed and dissatisfied. Just make sure that you gain a firm grip on your own field.
6. A lot of time is lost by looking for misplaced notes, books, journals, and reports. Make an unbreakable rule: a place for everything, and everything in its place.
7. "Most people," says motivational speaker Anthony Robbins, "would not see an awful movie a second time, yet they play the same bad memories over and over again in their heads."
8. When you're really through studying, spend an extra fifteen minutes studying just an extra bit more.

9. If the thought of saving time sounds sensible but uninspiring, ask yourself this simple question: What do I want to save time for? Suddenly, the efficient use of time may take on a significance that it never had before.

10. A Sanskrit proverb puts everything in proper perspective:

Today well lived
Makes every yesterday a dream of happiness
And every tomorrow a vision of hope
Look well therefore to this day.

HAVE YOU MISSED SOMETHING?

SENTENCE COMPLETION

Complete the following sentences with one of the three words listed below each sentence.

1. A time schedule functions as a ____________________.

   reward **game plan** **punishment**

2. When it comes to studying, "prime time" is usually ____________________.

   daytime **mid-afternoon** **flexible**

3. A master schedule should normally be drawn up once a ________________.

   day **week** **term**

MATCHING

In each blank space in the left column, write the letter preceding the phrase in the right column that matches the left item best.

_____	1. Pocket work	a. Work expands to fit the time allotted
_____	2. E-mail	b. Consumes a large portion of a typical first-year student's weekday
_____	3. Master schedule	c. Nobel Prize winner who used sleep time to solve problems
_____	4. "Free time"	d. Provides a basic structure for organizing the term's activities
_____	5. Pareto Principle	e. Rhythms that influence your body's cycle of sleeping and waking
_____	6. Szent-Györgyi	f. Productivity boon that can sometimes be a "time sink"
_____	7. Parkinson's Law	g. Supplies study material for unexpected free time
_____	8. Circadian	h. Also known as the 80/20 rule

TRUE-FALSE

Write T *beside the* true *statements and* F *beside the* false *statements.*

_____ 1. Taking breaks has a detrimental effect on comprehension.

_____ 2. Free time can often occur unexpectedly.

_____ 3. Using a time schedule can make you a slave to time.

_____ 4. Scheduling saves time that might otherwise be wasted.

_____ 5. An hour of daytime work is usually more productive than an hour at night.

MULTIPLE CHOICE

Choose the word or phrase that completes each sentence most accurately, and circle the letter that precedes it.

1. When your obligations and assignments begin to seem overwhelming, it helps to
   a. take a nap.
   b. divide your time into blocks.
   c. use Parkinson's Law.
   d. all of the above.

2. "Hidden" time is time that you
   a. aren't allowed to use.
   b. managed to overlook.
   c. reserve for recreation.
   d. leave off your schedule.

3. A good way to take inventory of your time habits is by
   a. checking your master schedule.
   b. asking an instructor to help you.
   c. keeping a daily activities log.
   d. none of the above.

4. The master schedule provides
   a. an alternative to a weekly schedule.
   b. an excuse for increased recreation.
   c. a basic structure for organizing your activities.
   d. a solution to the problem of hidden time.

5. A task-based schedule may be your best bet if
   a. your days are unpredictable.
   b. your free time is fluid.
   c. you have a full-time job.
   d. all of the above.

SHORT ANSWER

Supply a brief answer for each of the following items.

1. What is the purpose of keeping a daily activities log?
2. In what ways does a schedule promote relaxation?
3. Explain the possible role of a note pad in controlling your time.
4. Discuss how to prepare a task-oriented master schedule.

WORDS IN CONTEXT

From the three choices beside each numbered item, select the one that most nearly expresses the meaning of the italicized word in the quote. Make a light check mark (✓) next to your choice.

Mathematics has given economics *rigor,* but *alas,* also *mortis.*

—Robert L. Heilbroner (1919–), American economist

1. *rigor*	strength	exactitude	precision
2. *alas*	sorrowfully	interestingly	happily
3. *mortis*	humanity	fear	death

Interest you owe works night and day, in fair weather and foul. Interest gnaws at a man's *substance* with invisible teeth.

—Henry Ward Beecher (1813–1887), newspaper editor and clergyman

4. man's *substance*	brain	worth	body

Of all the mysteries of the stock exchange, there is none so *impenetrable* as why there should be a buyer for everyone who seeks to sell.

—John Kenneth Galbraith (1908–), Canadian-born American economist, writer, and diplomat (ambassador to India)

5. *impenetrable*	incredible	inexplicable	important

"Involvement" in this context differs from "commitment" in the same sense as the pig's and the chicken's roles in one's breakfast of ham and eggs. The chicken was *involved*—the pig was *committed.*

—Anonymous

6. chicken was *involved*	primed	rejected	associated
7. pig was *committed*	solicited	included	affiliated

THE WORD HISTORY SYSTEM

Tally
a reminder of the early method of counting

tally tal'-ly *n.* 1. A reckoning or score. 2. A mark used in recording a number of acts or objects.

Tally goes back to the time when things were commonly counted by cutting notches in a stick of wood. The word was borrowed in Middle English as *taille,* from Old French *taille,* "a cutting." It was formerly customary for traders to have two sticks and to mark with notches on each the number or quantity of goods delivered, the seller keeping one stick and the purchaser the other. When such records came to be kept on paper, the same word was used for them; and it now means almost any kind of count or score.

Reprinted by permission. From *Picturesque Word Origins* © 1933 by G. & C. Merriam Co. (now Merriam-Webster, Incorporated).

PART
II

Note-Taking Skills

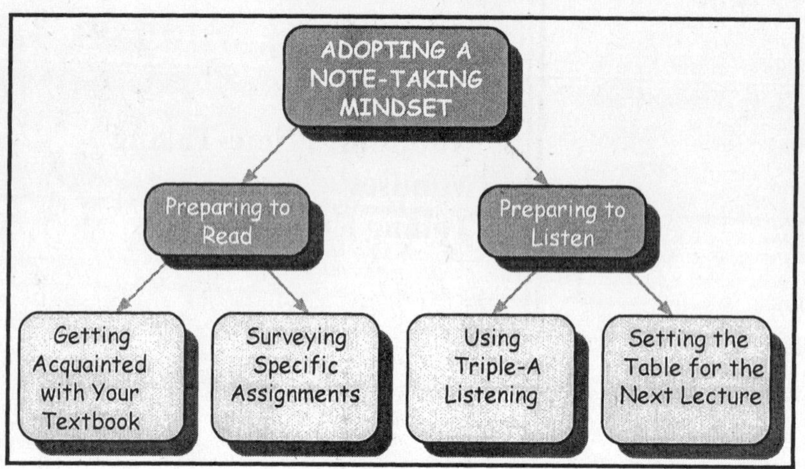
ADOPTING A NOTE-TAKING MINDSET
Preparing to Read
Preparing to Listen
Getting Acquainted with Your Textbook
Surveying Specific Assignments
Using Triple-A Listening
Setting the Table for the Next Lecture

CHAPTER
2

Adopting a Note-Taking Mindset

One important key to success is self-confidence. An important key to self-confidence is preparation.

Arthur Ashe
(1943–1993),
American tennis champion

Begin a match or run a race without warming up first, and you won't be doing yourself much good. You may even be doing some harm. It makes sense to warm up, psych up, and just plain be ready before you undertake anything of importance. Note taking is certainly no exception. This chapter talks about:

- **Preparing to Read**
- **Preparing to Listen**
- **Surveying Specific Assignments**
- **Setting the Table for the Next Lecture**

How do you give your note taking real value?

Note taking doesn't work in isolation. You need to build a background and create a context in order for the sentences you jot down or the paragraphs you mark up to have real value. If you don't speak the language of your textbook or know the way to listen effectively to a lecture, it's unlikely that you'll be in a position to take meaningful notes. And even if you do, you still need to take your preparations one step further by previewing each textbook assignment before you begin reading or by setting the table for an upcoming lecture. The time you invest at the beginning will strengthen your memories throughout and make you a much better student in the end.

Preparing to Read

What's the best way to prepare for reading?

Because you and your textbooks are going to spend a quarter or a semester together, you'd better become friends. How? By getting acquainted with the textbook as a whole and by surveying specific assignments as they arise.

Getting Acquainted with Your Textbook

What's the first thing you should do when you get a new textbook?

Not all textbooks are alike. Most have distinct styles of organizing their ideas and communicating their information. It's important to learn a textbook's language before you begin trying to understand what it has to say.

What steps can you take to become acquainted with a textbook?

Give yourself enough time to understand how each of your textbooks works by buying them as soon as you've registered. This is a wise policy even if your school allows a period in which you can attend many courses before deciding on a final few. Don't wait until the first day of class. Get a head start by reading the tables of contents, prefaces, introductions, and other up-front material in all your books. Underline important words and sentences, and make notes in the margins. Look at the pictures, tables, and diagrams; read the captions. Read chapter titles, headings, and subheadings that interest you. This will give you a good idea of what the book is like and where you will be going during the semester. Later, you'll be glad you did, for you'll be able to see how the various parts of the course fit together.

Read the Prefatory Material

How does a book's prefatory material differ from the rest of its pages?

Although most textbooks are written in a serious, scholarly tone, in the prefatory material—which may be called "Preface," "To the Student," or something similar—the authors often take a more personal approach to their subjects and their readers than they do in the body of the text. This provides an opportunity to meet the authors as people and to get comfortable with them. Once you do, you'll find that you can converse and even argue with them as you read the text. Now and then, you'll find yourself saying, "No, I don't agree with that statement," or "What do you mean by that?"

What sort of valuable information does a preface typically contain?

In the prefatory material you can also find valuable information about the concepts and content of the book. For instance, if you began this book by reading the "To The Student" section, you know the purpose and use of the questions that appear in the margins of every chapter. If, on the other hand, you skipped over this part, you may have squandered a valuable component of the book. In prefaces you can find valuable information such as (1) what the author's objective is and is not, (2) the organizational plan of the book, (3) how and why the book is different from other books about the same subject, (4) the author's qualifications for writing the book, and often (5) an explanation of any supplementary learning aids. As a practical exercise, you might find it interesting to read the preface of this book, if you have not already done so. See how much you can gain toward understanding not only the book but also the authors.

How does learning the author's objective help?

The Author's Objective Learning the author's objective—the purpose or goal he or she meant to achieve—enables you to read and interpret the text appropriately. For instance, the authors of *American Government* state their objective succinctly:

> In preparing every edition of *American Government,* including this new ninth edition, we have committed ourselves to two major goals: (1) to write a concise yet comprehensive textbook that (2) helps students think critically about the U.S. political system.[1]

In a single sentence, they tell you straight off what to expect and what particular philosophy they will be following.

---

[1] Alan R. Gitelson, Robert L. Dudley, and Melvin J. Dubnick, *American Government,* 6th ed. (Boston: Houghton Mifflin, 2001), p. xv.

Why is it helpful to know a book's organizational plan?

The Organizational Plan of the Book Knowing the book's organizational plan is like having a road map. You know not only what the authors are doing but also where they are going. In the following example, you learn of the carefully organized introduction of topics as well as the self-contained, flexible nature of the individual chapters:

> The contents have been arranged so that as each new topic is introduced, it is fully explained and its fundamentals are thoroughly examined before commencing further study. For example, Chapter 13, Investment Fundamentals and Portfolio Management, precedes chapters on specific types of investments. In addition, not only does each chapter follow an overall logical sequence, but each is also a complete entity. The chapters can therefore be rearranged to follow any instructor's developmental sequence without losing students' comprehension.[2]

What is the advantage of knowing why a book is unique?

How and Why the Book Is Different Recognizing what makes the textbook unique enables you to read with greater awareness and comprehension and to avoid the trap of thinking the book is "more of the same old stuff." In this case, the author chooses to reinforce the unique nature of his textbook by placing this paragraph in boldface:

> The text you are holding in your hands is literally the product of the research program undertaken in our laboratory. I have had opportunities to present these data at professional conferences and am encouraged by the responses I have received from course instructors. As one reviewer of the book put it, "Finally there is a text that uses psychology to teach psychology."[3]

Why do writers want you to know they are experts?

The Author's Qualifications Writers usually try in some subtle way to let the reader know that their book is written by an expert on the subject and that therefore the information is trustworthy and credible. Note in the following example how the authors allude to their "considerable experience" while explaining their book's philosophy:

---

[2] E. Thomas Garman and Raymond E. Forgue, *Personal Finance*, 7th ed. (Boston: Houghton Mifflin, 2003), p. xxviii.

[3] Jeffrey S. Nevid, *Psychology: Concepts and Applications* (Boston: Houghton Mifflin, 2003), p. xviii.

> It is our conviction, based on considerable experience in introducing large numbers of students to the broad sweep of civilization, that a book reflecting current trends can excite readers and inspire a renewed interest in history and the human experience.[4]

What does the prefatory material tell you about learning aids?

Supplementary Learning Aids Today's textbook is often one component of a larger package of learning aids. The prefatory material generally names these aids and tells how they'll benefit you. In addition to in-text features, supplementary materials such as workbooks, study guides, software, Web sites, and videos are often available. Reading the introductory material will alert you to the presence of these features and ancillary materials, as you can see from the following example:

> The teaching package for the fifth edition includes the *Instructor's Resource Manual with Test Questions, Case Teaching Guide,* a computerized test bank, a videotape with instructor's guide, PowerPoint slides, and student and instructor web sites, which are new to this edition.[5]

Read the Introduction

What is the role of a book's introduction?

While the prefatory material often supplies a glimpse of the author's personality, the introduction provides a preview of the book itself. The introduction is the book's showcase. It is designed not only to introduce a book, but also to sell it. Authors and publishers both know that this is the place prospective customers often go to decide whether a book is worth reading and worth buying. As a result, the introduction is often especially well written and inviting.

Figure 2.1 is a densely packed introduction containing information of great and immediate value for the sharper reading of textbooks. It is from a book titled *Six-Way Paragraphs.*[6] The book's sole purpose is to teach students how to spot main ideas. One hundred paragraphs are provided for practice. To prepare students for such practice, the introduction strives to explain the ins and outs of the paragraphs found in textbooks. As you read it, be aware not only of *what* the writer says but also of *how* he says it and his *purpose* for saying it.

---

4 John P. McKay, Bennett D. Hill, John Buckler, and Patricia Buckley Ebrey, *A History of World Societies,* 5th ed. (Boston: Houghton Mifflin, 2000), p. xxvii.

5 Jean-Pierre Jeannet and H. David Hennessey, *Global Marketing Strategies,* 5th ed. (Boston: Houghton Mifflin, 2001), p. xviii.

6 Walter Pauk, *Six-Way Paragraphs* (Providence, RI: Jamestown Publishers, 1974).

what: wants you to focus on the paragraph-unit how: brings you and the writer together purpose: wants you to look at the paragraph through the eyes of the writer	The paragraph! That's the working-unit of both writer and reader. The writer works hard to put meaning into the paragraph; the reader works hard to take meaning out of it. Though they work at opposite tasks, the work of each is closely related. Actually, to understand better the job of the reader, one must first understand better the job of the writer. So, let us look briefly at the writer's job.
what: each paragraph has but one main idea how: shows you how a writer thinks purpose: to convince you to look for only one idea per paragraph because writers follow this rule	To make his meaning clear, a writer knows that he must follow certain basic principles. First, he knows that he must develop only one main idea per paragraph. This principle is so important that he knows it backward, too. He knows that he must not try to develop two main ideas in the same, single paragraph.
what: the topic of the main idea is in the topic sentence, which is usually the first one how: the writer needs to state a topic sentence to keep his own writing clear and under control purpose: to instill confidence in you that the topic sentence is an important tool in a writer's kit and convince you it is there, so, look for it!	The next important principle he knows is that the topic of each main idea must be stated in a topic sentence and that such a sentence best serves its function by coming at or near the beginning of its paragraph. He knows, too, that the more clearly he can state the topic of his paragraph in an opening sentence, the more effective he will be in developing a meaningful, well-organized paragraph.
what: developing main ideas through supporting material how: "more to a writer's job," still keeps you in the writer's shoes purpose: to announce and advance the new step of supporting materials	Now, there is more to a writer's job than just writing paragraphs consisting of only bare topic sentences and main ideas. The balance of his job deals with *developing* each main idea through the use of supporting material that amplifies and clarifies the main idea and many times makes it more vivd and memorable.

Figure 2.1
The Content of an Introduction

Source: Walter Pauk, *Six-Way Paragraphs* (Providence, RI: Jamestown Publishers, 1974), pp. 7–8. Reprinted by permission.

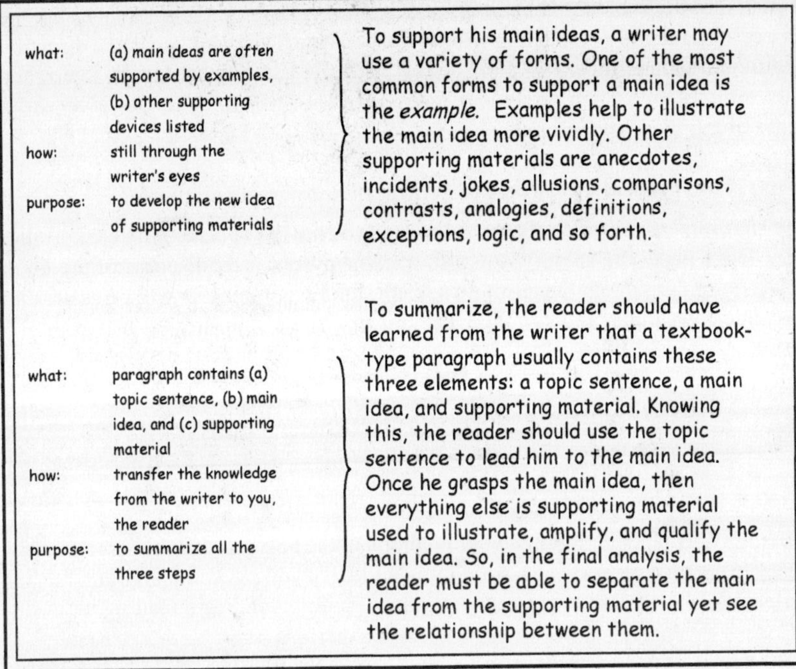

what:	(a) main ideas are often supported by examples, (b) other supporting devices listed	To support his main ideas, a writer may use a variety of forms. One of the most common forms to support a main idea is the *example.* Examples help to illustrate the main idea more vividly. Other supporting materials are anecdotes, incidents, jokes, allusions, comparisons, contrasts, analogies, definitions, exceptions, logic, and so forth.
how:	still through the writer's eyes	
purpose:	to develop the new idea of supporting materials	
what:	paragraph contains (a) topic sentence, (b) main idea, and (c) supporting material	To summarize, the reader should have learned from the writer that a textbook-type paragraph usually contains these three elements: a topic sentence, a main idea, and supporting material. Knowing this, the reader should use the topic sentence to lead him to the main idea. Once he grasps the main idea, then everything else is supporting material used to illustrate, amplify, and qualify the main idea. So, in the final analysis, the reader must be able to separate the main idea from the supporting material yet see the relationship between them.
how:	transfer the knowledge from the writer to you, the reader	
purpose:	to summarize all the three steps	

Figure 2.1
The Content of an Introduction—*continued*

Preview from Front to Back

What can you learn from a book's table of contents?

After you've read the introductory material, survey the rest of the book. Begin by scanning the table of contents (TOC). It lists the parts, the chapters, and sometimes the major headings within each chapter. Some books actually have two TOCs, an abbreviated one that supplies the part and chapter names and a more detailed table of contents that lists the names and page references of sections and subsections in addition to those for the chapters and parts. General or detailed, the TOC is designed to show the overall organization of the book and how the chapter topics relate to one another. It also indicates whether the book contains extra material: appendixes, glossaries, bibliographies or references, and indexes.

What sort of valuable information can you glean from a book's extra sections?

Now turn to the back of the book and look at these extra sections. Appendixes contain additional information such as tables and graphs, documents, or details about specific aspects of a subject. Glossaries are specialized dictionaries of terms common to the subject the book discusses. Bibliographies and references list the sources the authors consulted in writing the book and can point you in the direction of further reading. Indexes—alphabetical listings of significant topics, ideas, and names along with page references that appear at the end of texts—give you a sense of the scope of the book and help you locate specific material quickly.

What is the general advantage of getting to know your textbook?

By familiarizing yourself with your textbook, you not only bolster your background knowledge about the subject; you also become aware of the features of the book you can use throughout the term. As a result, future assignments will be easier and less time consuming, and you'll have a greater chance to master the material.

Surveying Specific Assignments

What is the value of surveying a reading assignment?

Moving systematically through your textbook at the beginning of the semester will give you a strong context and a solid foundation that you can strengthen and further build on by briefly surveying your assignment before you begin reading it.

Why Surveying Is Important

How important is it to survey your assignment first?

A good scholar would no more begin reading a chapter without first skimming it than an automotive engineer would run a car without first greasing it. The grease does not supply the power, but without it the gasoline would not be of much use. If you try to cut corners by skipping the skimming, you will lose time, not save it. If you burrow directly into one paragraph after another, you'll be unearthing one compartmentalized fact after another, but you won't see how those facts relate to each other. Psychologists call this not-seeing-the-big-picture condition *tunnel vision.*

How does surveying make a difference in your reading?

There are three practical ways that surveying can make a real and immediate difference in your reading: by creating a background, by limbering your mind, and by overcoming inertia.

How does surveying create a background?

1. *Surveying creates a background.* When you don't have some prior knowledge about the subject matter of an assigned chapter, you read slowly and have difficulty understanding the material. When you come to something that you

recognize, your reading speed quickens and your comprehension grows. The difference is your prior knowledge. Surveying counteracts tunnel vision and prepares you for reading by giving you some background information about a chapter.

What are advanced organizers?

According to David P. Ausubel, a learning-theory psychologist, previewing the general content of a chapter creates advance organizers, which help students learn and remember material they later study closely.[7] These familiar landmarks act as topics or categories under which ideas, facts, and details may be clustered. John Livingston Lowes, a professor of literature at Princeton University, characterized such familiar landmarks as *magnetic centers* around which ideas, facts, and details cluster like iron filings around a magnet.

What did Katona's test find about advance organizers?

George Katona, a psychologist, tested the effectiveness of advance organizers with two groups of students. One group was asked to read a selection in which a general principle of economics was stated in the first sentence. The second group was given the same selection, but with the first sentence deleted. When the students in the first group were tested, they not only remembered the specific content of the paragraph better than the second group, but they were able to apply the general principle to all the examples in the selection. Without the first sentence, which was an advance organizer, students in the second group viewed the examples as separate, unrelated entities and were unable to see that they could be clustered under the umbrella of a common principle in economics.

What are some examples of the way in which surveying creates a background?

When you skim a chapter, you spot and pick up topics by reading headings and subheadings. You pick up ideas by reading the first and last sentences of paragraphs. You become familiar with the names of people and places by skimming these names. You grasp the general objective of the chapter by reading the introductory paragraph, and you get an overview by reading the summarizing paragraph at the end of the chapter. All of these elements combine to provide the background you need for strengthening your memory and serve as a foundation for learning that lasts. Of course, it's unlikely that you'll know any of these facts and ideas cold after the first go-round. But when you encounter them again during your careful reading, you will recognize them, and this familiarity will give you confidence and understanding.

---

[7] John F. Wakefield, *Educational Psychology* (Boston: Houghton Mifflin, 1996), p. 368.

How does surveying compare to an athlete's pregame warm-up?

2. *Surveying limbers the mind.* For an athlete, a pregame warm-up limbers the muscles as well as the psyche and brain. An athlete knows that success comes from the coordination of smoothly gliding muscles, a positive attitude, and a concentrating mind. The prestudy survey of a textbook achieves for the scholar what the pregame warm-up achieves for the athlete.

How does surveying overcome inertia?

3. *Surveying overcomes mental inertia.* How often have you said, with impatience and exasperation, "C'mon. Let's get started!" Getting started is hard. According to Newton's First Law of Motion (also known as the Law of Inertia), "A body in motion tends to remain in motion; a body at rest tends to remain at rest." Many students find it difficult to open a textbook and begin to study. If you are one of them, use surveying to ease yourself into studying. Surveying does the job: It gets you started.

Do you need to survey the entire chapter before you begin reading?

Although surveying is essential in successfully completing a textbook assignment, you need not always survey an entire chapter as the first step. You may begin by surveying the first part before you read it. Later, as you work your way through the chapter, you may want to skim further ahead, page by page, as you read and study to understand. Whether you survey your assignment all at once or in chunks, it's a step you don't want to skip.

How to Survey an Assignment

What is the advantage of previewing your assignments?

Surveying specific assignments allows you to overcome inertia, delve into the text, and get a sense of the larger picture into which you can fit specific ideas and facts. When previewing, you can use any or all of the following techniques.

How should you think about a chapter's title?

Think About the Title Take a few moments to reflect on the chapter's title. Is its meaning obvious? If not, can you guess what the chapter will discuss?

What is the value of reading the introduction and summary first?

Read the Introduction and the Summary Although they may not be marked as such, textbook chapters frequently include an introductory paragraph, a brief summary at the end, or both. Reading these sections provides advance organizers, which create a framework on which you can build the information in the chapter.

What role do headings and subheadings play in previewing?

Look Over Headings and Subheadings Flip through the pages and read the headings and subheadings. Notice their hierarchy and sequence. This will provide a big-picture view of where the chapter will take you.

Why should you pay attention to information set apart from the rest of the text?

Note Any Information Set Apart from the Rest of the Text In general, information that is boxed, boldfaced, bulleted (preceded with dots or squares), screened (printed on a gray or colored background), or otherwise set apart is information that you don't want to miss. Take a moment to look it over now so you will already be familiar with it when you go back to read the chapter completely.

How do visuals help your understanding of a chapter assignment?

Glance at the Visuals Pictures and graphic materials can provide a distillation of an entire chapter's ideas in less than a page. They can sometimes succinctly express in a small space what may take several paragraphs to explain in words. The concept maps at the start of each chapter in this book, for example, show the chapter at a glance. Other visuals supply vivid and easy-to-understand examples of key points.

Is there a downside to previewing a textbook assignment?

As important and helpful as it is to preview a specific reading assignment, you may still want to limit the amount of time you spend surveying the chapter. After all, your primary purpose is to read and understand the assignment. If you spend too much time preparing to read, you'll run out of time and energy for the actual reading. Or as an old Chinese proverb warns, "Keep sharpening your knife and it will grow dull."

Preparing to Listen

How do you develop good listening habits?

Of the many skills a person needs to learn in this complex world, listening is just about the easiest to acquire. And yet it is a skill that is often overlooked. Figure 2.2 presents a short quiz to help you learn where you stand as a listener. All you need to do is recognize the habits of a poor listener as well as the techniques used by a good listener. Then, knowing the good techniques, apply them without exception in all your daily activities. You can make a fresh and immediate start today by referring to Figure 2.3, which lists the ten most important keys to effective listening.

How often do you find yourself engaging in these ten bad habits of listening? Check the appropriate columns.

How often do you ...	Frequency: Almost always	Usually	Sometimes	Seldom	Almost never	Score
Decide that the topic is boring?						
Criticize the speaker?						
Overreact by disagreeing?						
Listen only for bare facts?						
Outline everything?						
Fake attention?						
Yield to distractions?						
Avoid listening to tough technical information?						
Let emotion-laden words arouse personal antagonism?						
Waste thought speed by daydreaming?						
Total score						

Tally your score as follow:

Almost always	2
Usually	4
Sometimes	6
Seldom	8
Almost never	10

Interpret your score as follows:

Below 70	Need training in listening
70 to 90	You listen well
Above 90	Extraordinarily good listener

Figure 2.2
Listening Habits

Source: Adapted from Ralph G. Nichols and Thomas R. Lewis, *Listening and Speaking,* p. 166. Originally published by Wm. C. Brown Group, 1954. Reprinted by permission of the authors.

Keys to Effective Listening	The Poor Listener	The Good Listener
1. Find areas of interest	Tunes out dry topics.	Seizes opportunities: "What's in it for me?"
2. Judge content, not delivery	Tunes out if delivery is poor.	Judges content, skips over delivery errors.
3. Hold your fire	Tends to enter into argument.	Doesn't judge until comprehension is complete.
4. Listen for ideas	Listen for facts.	Listens for central themes.
5. Be a flexible note taker	Is busy with form, misses content.	Adjusts to topic and organizational pattern.
6. Work at listening	Shows no energy output, fakes attention.	Works hard; exhibits alertness.
7. Resist distractions	Is distracted easily.	Fights or avoids distractions; tolerates bad habits in others; knows how to concentrate.
8. Exercise your mind	Resists difficult expository material; seeks light, recreational material.	Uses heavier material as exercise for the mind.
9. Keep your mind open	Reacts to emotional words.	Interprets emotional words; does not get hung up on them.
10. Thought is faster than speech; use it	Tends to daydream with slow speakers.	Challenges, anticipates, mentally summarizes, weighs the evidence, listens between the lines to tone of voice.

Figure 2.3
Ten Keys to Effective Listening

Source: Reprinted by permission of Unisys Corporation.

Using Triple-A Listening

What are the three basic skills of good listening?

Listening is the first step in lecture note taking. Being a good listener means being an active listener. But listening is not the same as hearing. Listening isn't simply a matter of acting as a human microphone and picking up the sound of the lecturer's voice. It is a conscious activity based on three basic skills: *a*ttitude, *a*ttention, and *a*djustment. These skills are known collectively as *triple-A listening.*

A_1: Maintain a Positive Attitude

Why do you need a positive attitude to listen effectively?

You can take many steps to improve your listening, but the primary prerequisite to effective listening is a positive mental attitude. A positive attitude sets the stage for the open-mindedness that is essential to learning and comprehension. As businessman and author Kevin J. Murphy says in *Effective Listening,* "Minds are like parachutes; they only function when open."[8] You must walk into the classroom convinced that the lecturer has something useful to say. If you have any doubts, take a moment to think about the kind of preparation that goes into a typical lecture. The lecturer had to do the searching, reading, selecting, discarding, and organizing of information from dozens of books and other sources—a task that can take hundreds of hours. You can reap the benefits of all that effort in a single lecture.

What can you do to help cultivate a positive attitude?

Cultivating a positive attitude isn't always easy. But it is worth the effort. If you're struggling to give the speaker the benefit of the doubt, try finding areas of interest in what he or she has to say, judge the content of the lecture instead of the delivery, and hold your fire if you're inclined to lash out at those things you don't agree with.

How can you find areas of interest in a "boring" lecture?

Finding Areas of Interest It doesn't take any talent to tune out a lecture or to brand it as "boring." Anyone can do that. Most speakers work hard to make their lectures informative and interesting. But they can't do it alone. It requires some effort on your part as well. Don't waste your time by ignoring the instructor or by reading your textbook when you should be giving the speaker your full attention. You can usually find an element or an aspect of the lecture topic that appeals to your personal interests. For example, a student who played basketball developed a greater interest in physics when he realized the instructor's discussions of trajectory and momentum could be applied to his foul shots.

What if you can't find anything in the lecture that interests you?

Even if you can't immediately find something about the lecture that interests you, it still makes sense to listen intently, take notes vigorously, and show interest by your facial expression. These outward manifestations of interest will help generate genuine internal interest.

---

8 Kevin J. Murphy, *Effective Listening: Hearing What People Say and Making It Work for You* (New York: Bantam Books, 1987), p. 28.

What is the most important aspect of a lecture?

Judging Content, Not Delivery Listeners are sometimes terribly rude or downright cruel when a lecturer's delivery fails to measure up to some preconceived standard. Keep in mind that great thinkers aren't always excellent orators. Try to approach each lecture with a humane attitude, and concentrate on what the lecturer is saying, not on how it's being said. If you do, you'll emerge from the lecture with a positive self-image. You will also learn something.

What should you do if the lecturer says something you don't agree with?

Holding Your Fire In college you're bound to hear ideas that are different from or even contrary to those you hold. When this happens, your knee-jerk reaction may be to speak up immediately to defend your position. Hold your fire! Try to listen intently to thoroughly understand the viewpoint expressed by the lecturer. If you shift all your focus to rebutting a point you don't agree with, you'll be unable to concentrate on the rest of the lecture. In a war of words of this type, understanding is often the first casualty.

A_2: Strive to Pay Attention

Why is attention important?

Attention is the path that will lead you into the wonderful state of concentration. Without concentration there is no focus, and without focus there is almost no learning. If you pay attention and concentrate, you will become an active listener able to synthesize new information with facts and ideas you've already known.

How can you build anticipation?

Attention often thrives on anticipation. Before class, look over your notes from the last lecture, and take a minute to speculate about what your instructor is going to talk about today. If the lectures follow your textbook, peek ahead to see what's coming. Once the lecture starts, let your mind dart ahead (during pauses) to anticipate what's coming next. You'll be alert, engrossed in the material, and concentrating 100 percent.

What should your attention be focused on?

Of course, you cannot attain concentration by concentrating on concentration. Your attention must focus on the lecture, not on the act of concentration itself. Deep cognition, or deep thinking, is vital. When you hear a lecture, the words enter your short-term memory, where they have to be swiftly processed into ideas. Active listening sets that process in motion. If your instructor's words are not processed through the kind of active listening that attention sets in motion, they will be dumped from short-term memory and gone for good. If you process the words into ideas, in a flash the ideas will be stored in your long-term memory.

How can you foster the kind of attention that leads to concentration?

You can foster the kind of attention that leads to concentration and in turn to active listening by focusing on ideas instead of facts, by working to stay fully engaged, by tuning out distractions, and by using the natural lag between speech and thought to forge ahead in your mind in order to consider a speaker's current point and anticipate the next one.

What's the drawback of listening for facts instead of ideas?

Listening for Ideas Don't imitate the detective who says, "The facts, ma'am, just stick to the facts." Sure, facts are important. And it's true that they're often what you'll be tested on. But facts in isolation don't hold together well. In order to organize a lecture's information in both your notes and your memory, look to the ideas that lie behind the facts. Strive to uncover the principle or idea that each fact is supporting. Try to see how the pieces of the puzzle fit into the big picture.

How do you work at listening?

Working at Listening Listening isn't a passive activity; it requires true participation. Good listeners are fully engaged, outwardly calm, but inwardly dynamic; and they sit toward the front of the classroom. While taking notes, the listener may nod in agreement or look quizzical when the presentation becomes unclear. These sorts of gestures aren't just for effect. Such activity actually promotes comprehension and learning by the listener and provides encouragement to the speaker.

What's the best way to resist distractions?

Resisting Distractions There are distractions aplenty in the classroom: antics of other students, whispering, the speaker's dress and mannerisms, outside noise, and outside views. In some classes, students may actually be browsing the Web or sending instant messages while the lecturer is speaking! The best way to resist distractions and maintain your concentration is to rivet your eyes on the speaker when you have a chance and focus on taking notes the rest of the time. You may also want to consider sitting close to the front. The level of distraction tends to increase as you move farther away from the instructor, especially in large lecture halls.

How can you take advantage of thought speed?

Using Your Thought Speed In a race of ideas, thought is speedier than speech. Most of us can grasp a point faster than a lecturer can deliver it. This "thought speed" leaves a bit of a lag during which your mind is apt to wander. Rather than using this time to daydream or to dart off on mental side trips, devote it to thinking more deeply about what the lecturer is saying. Mentally enumerate the ideas that have been expressed and summarize them until it's time to shift your focus back to what's being said. Keep alternating in this fashion throughout the lecture.

A_3: Cultivate a Capacity for Adjustment

Why is adjustment so important to active listening?

Even though most speakers try to indicate what they intend to cover in their lectures, you still need to be mentally limber enough to follow a lecture regardless of the direction it may take. Sometimes a speaker says, "This event had three important results," but then discusses four or five. Other times a question from the audience suddenly shifts the speaker's focus. In such cases, you can't simply tune out the parts of the lecture that don't fit with your expectations. You have to be able to "roll with the punches." That's why adjustment is such an important component of active listening. You can cultivate your capacity for adjustment by remaining flexible while taking notes, by exercising your mind by sitting in on subjects that you are unfamiliar with, and by keeping your mind open to ideas and information that you may initially want to tune out.

What is the key to being a flexible note taker?

Being a Flexible Note Taker Flexibility in note taking depends on informed listening. Informed listening means being able to identify the organizational patterns used by the speaker. Organizational patterns are easily recognizable if you know their basic structures in advance. Once you have detected a pattern, you'll have a framework you can use to anticipate the form of the message and adjust the way you record your notes. Similarly, when the speaker shifts gears and adopts a different pattern, a flexible note taker should be able to adjust accordingly.

How can you exercise your mind to become a better listener?

Exercising Your Mind From time to time, try to sit in on lectures in fields that you know very little about. When you do, make a serious effort to follow the lecturer's chain of thoughts and ideas. True, some parts of the lecture may be unintelligible to you, but you're bound to understand other parts of it. Such listening can be hard work, but in the same way that hard work at a gym can strengthen your muscles, the hard work of listening will strengthen your will to concentrate and your power to persist.

How can you prevent highly charged words from interfering with your ability to listen?

Keeping Your Mind Open It is hard to believe that a single word or phrase can cause an emotional eruption. Among poor listeners, however, that is frequently the case, and even among very good listeners fireworks occasionally go off. Words that are red flags for some listeners include *activist, liberal, conservative, evolution, creationism, feminist, abortion, pro-life,* and *free market.* Dealing with highly charged language is similar to dealing with a fear or phobia. Often, the emotional impact of the words can be decreased through a frank and open discussion.

Don't be so inflexible as to totally shut out opposing ideas, as incendiary as they may seem at first. Genuinely listen to the other point of view. You might learn something that you didn't know before.

Setting the Table for the Next Lecture

What information can you use to help prepare for an upcoming lecture?

Once you've cultivated listening skills to build an effective base for comprehending any lecture, you can zero in on the upcoming lecture by marshalling the available information to provide a context for what the speaker is going to talk about. Of course, lectures aren't like textbook chapters; unless they've been written down or recorded already, you can't survey them in advance. But in most courses, there is enough available information supporting the lecture that it should enable you to effectively prepare for it. Each lecture can be viewed as a piece of a puzzle. The more pieces you have in place, the more you know about the shape and size of the pieces that remain. The course syllabus, the notes from your last lecture, and related reading assignments can all function as these puzzle pieces as you prepare for a lecture.

Look Over the Syllabus

How can you use your course syllabus to prepare for an upcoming lecture?

Assuming that your instructor is sticking to the syllabus, you should be able to read the list of lecture topics to get a big-picture sense of where this latest lecture will fit in. Pay especially close attention to the title and description of the previous lecture, the upcoming lecture, and the one that will follow. Encourage active thinking about the lecture topics by transforming each title into a question and then answering your question in cases where the lecture has already been delivered or by speculating on the answer in cases where the lecture is still to come.

Review Your Notes from the Previous Lecture

Why is it helpful to prepare for an upcoming lecture by reviewing your notes from the previous lecture?

Each lecture will normally build on the concepts from the lectures that came before it. Most instructors will assume that you attended, understood, and can remember the previous lecture. After all, time will rarely permit a rehash of ideas that have already been fully explained. Although the upcoming lecture will often have some ties to all the lectures that came before it, the bond will usually be strongest with the most recent one. In fact, depending on the course, the current lecture may simply be a continuation of the previous lecture. In some courses, each lecture is like the next episode of a continuing story. To understand what's going on in Chapter 2 in a novel, you must understand and remember Chapter 1. Lectures often work in a similar way.

Do the Assigned Reading

How do reading assignments relate to the upcoming lecture?

A course syllabus will often include related reading assignments as well as lecture topics. Some instructors will assign readings as background for an upcoming lecture. Others will give you readings designed to follow up on what you've just heard and learned in a lecture. In either case, these readings can help provide advance organizers that will make it easier to understand and remember an upcoming lecture.

How can you prepare by using readings assigned before the lecture?

If a reading assignment precedes the lecture. Skipping the assigned reading can be as bad as skipping the previous lecture. You run the risk of sitting through an hour of discussion on topics that you are unfamiliar with or don't understand. If this happens, the time you spend in the classroom will largely be wasted. If the instructor assigns readings, make sure you do them. And once you've read your assignment, refresh your memory of what you've read by conducting a quick review shortly before the lecture begins. The concepts from the reading will provide advance organizers that will enable you to arrange and better remember what you learn in the lecture.

How can you prepare by using readings assigned after the lecture?

If a reading assignment follows the lecture. Most students are too busy to read ahead and complete assignments that aren't due. That's understandable. Furthermore, future reading assignments may deal with concepts that won't be clear until after you've attended the lecture. Even so, you should be able to skim an upcoming reading assignment without the knowledge and perspective you need to fully understand it. A quick survey of the assignment may prompt more questions than answers, but it will enable you to walk into a lecture with a pretty good idea of the sort of topics the speaker will be covering as well as a broader sense of where the course is heading.

FINAL WORDS

Why should you invest your time in preparing to take notes?

In a sense, note-taking preparation is a kind of investment. It requires you to do some work up-front that seems to take up time, yet yields few or no immediate results. But very few long-term benefits come without initial and sometimes costly investments. As a student, you understand this better than most. You appreciate the fact that the time and tuition you are devoting to your education will pay off in the long run. Apply this same sort of thinking to the preparations you make for taking notes. You'll be glad you did.

HAVE YOU MISSED SOMETHING?

SENTENCE COMPLETION

Complete the following sentences with one of the three words listed below each sentence.

1. Effective note taking requires building a background and creating a ________________.

   context **summary** **memory**

2. A good listener places great emphasis on the speaker's ________________.

   delivery **appearance** **ideas**

3. In listening to a lecture, attention often thrives on ____________________.

   appearance **anticipation** **agreement**

MATCHING

In each blank space in the left column, write the letter preceding the phrase in the right column that matches the left item best.

_____	1. Surveying	a.	May distract some listeners from the content of a lecture
_____	2. Attention	b.	Often reveals a textbook's overall organizational plan
_____	3. Preface	c.	Serves as a preview of what's to come in a textbook
_____	4. Background	d.	Can express a chapter's worth of ideas in a single page
_____	5. Visuals	e.	Sometimes faked by a poor listener
_____	6. Delivery	f.	Helps provide context for an upcoming lecture
_____	7. Thought	g.	Developed when you survey a textbook before reading it
_____	8. Syllabus	h.	Moves at a faster rate of speed than speech

TRUE-FALSE

Write T *beside the* true *statements and* F *beside the* false *statements.*

_____ 1. It is helpful to acquire your course textbooks as early as possible.

_____ 2. The first few pages of a textbook should usually be skipped.

_____ 3. Surveying a chapter provides advance organizers.

_____ 4. The terms *listening* and *hearing* are synonymous.

_____ 5. One way to exercise your mind is by sitting in on lectures in fields you know little about.

MULTIPLE CHOICE

Choose the word or phrase that completes each sentence most accurately, and circle the letter that precedes it.

1. A textbook's prefatory material is likely to discuss
   a. the author's objective.
   b. the book's organizational plan.
   c. supplementary materials.
   d. all of the above.

2. The prefatory material in a textbook does not include
   a. the book's purpose.
   b. the author's credentials.
   c. the book's organization.
   d. the course's objective.

3. The primary prerequisite to effective listening is
   a. a positive mental attitude.
   b. systematic note-taking skills.
   c. a thorough hearing examination.
   d. all of the above.

4. Listening is
   a. strictly mechanical.
   b. the same as hearing.
   c. not automatic.
   d. all of the above.

5. When you disagree with a lecturer's statement,
   a. begin rehearsing your rebuttal.
   b. automatically decide he or she is misinformed.
   c. take his or her stand as a bias.
   d. make a note of it, but keep on taking notes.

SHORT ANSWER

Supply a brief answer for each of the following items.

1. What are the advantages of buying your textbooks well before classes begin?
2. Outline some of the things you can expect to learn from a textbook's preface.
3. Which listening skill is the most important? Why?
4. Explain the greatest impediment to listening.

WORDS IN CONTEXT

From the three choices beside each numbered item, select the one that most nearly expresses the meaning of the italicized word in the quote. Make a light check mark (✓) next to your choice.

Failure! There is no such word in all the bright *lexicon* of speech, unless you yourself have written it there! There is no such thing as failure except to those who accept and believe in failure.

—Orison Swett Marden (1906–1975), American lawyer

1. *lexicon* of speech expressions idioms dictionary

In the depth of winter, I finally learned that within me there lay an *invincible* summer.

—Albert Camus (1913–1960), French novelist, essayist, and dramatist

2. an *invincible* summer fiery hot unbeatable intense

The gambling known as business looks with *austere* disfavor upon the business known as gambling.

—Ambrose Bierce (1842–1914), American author

3. with *austere* disfavor stern enormous particular

THE WORD HISTORY SYSTEM

Calculate
from the counting stones of Romans

calculate cal'-cu-late' *v.* 1. To ascertain by computation; reckon. 2. To make an estimate of; evaluate.

The Romans had no adding machines. Even the art of writing was known to comparatively few persons. So they did their adding and subtracting with the aid of little stones used as counters. The Latin word for the little stone used in this way was *calculus,* diminutive of *calx,* meaning "limestone." From *calculus* the verb *calculare,* "to calculate," was formed, and its past participle, *calculatus,* is the immediate origin of English *calculate.*

Reprinted by permission. From *Picturesque Word Origins* © 1933 by G. & C. Merriam Co. (now Merriam-Webster, Incorporated).

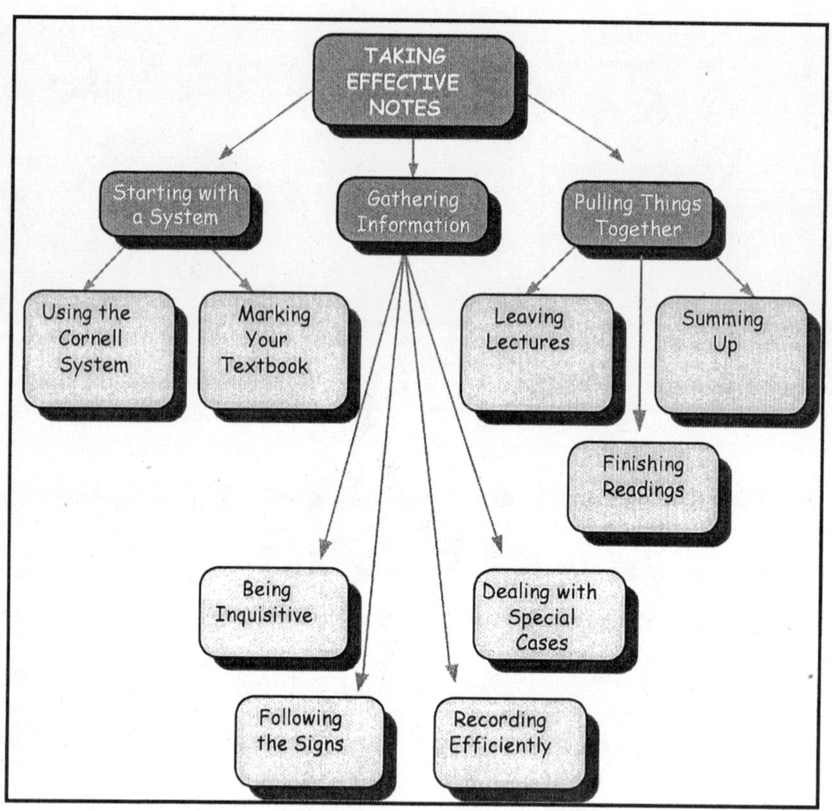
TAKING EFFECTIVE NOTES
Starting with a System
Gathering Information
Pulling Things Together
Using the Cornell System
Marking Your Textbook
Leaving Lectures
Summing Up
Finishing Readings
Being Inquisitive
Dealing with Special Cases
Following the Signs
Recording Efficiently

CHAPTER 3

Taking Effective Notes

Learn, compare, collect the facts!

Ivan Petrovich Pavlov
(1849–1936),
Russian physiologist

The lectures you hear and the assignments you read are a potential gold mine. Both include valuable nuggets of knowledge. Because you spend so much of your academic life accumulating information, it's unwise to do so haphazardly. After all, anybody can take notes, just as anyone can doodle. But it's crucial for your notes to go beyond mere busywork. The key is to take notes effectively. This chapter maps out a strategy for effective note taking:

- **Starting with a System**
- **Gathering Information**
- **Pulling Things Together**

How do you build on the framework of preparing for lectures and assignments?

Although it's obviously very important to have the right mindset before reading a textbook assignment or listening to a lecture, the efforts you make to prepare and foster a good frame of mind will be wasted if you don't build on this foundation. The way to do this is by taking notes.

What is the primary purpose of note taking?

The primary purpose of note taking is to provide a written record of what you've read or heard. You can't depend on short-term memory to retain all the ideas in a typical lecture or textbook chapter. Forgetting can be instantaneous and complete. For example, who hasn't forgotten a name only minutes after an introduction? Or had to reread a telephone number after getting a busy signal?

What does research tell us about the fragility of memory?

Given these experiences, it would be folly to rely solely on memory for something as important as an academic course. Carefully controlled research further points out the fragility of our memories. Experiments have shown that unrehearsed information is sometimes forgotten in as little as twenty seconds.[1] In a classical experiment, Hermann Ebbinghaus found that almost half of what is learned is forgotten within an hour.[2] Recently, psychologists carrying out experiments similar to Ebbinghaus's affirmed his findings.

The following true story further confirms the rapidity and scope of forgetting. Three professors eating lunch in the faculty lounge had this conversation:

CLYDE: Did you hear last night's lecture?
WALTER: No, I was busy.
CLYDE: Well, you missed one of the best lectures in recent years.
LEON: I agree. The four points that he developed were gems.
CLYDE: I never heard anyone make his points so clearly.
WALTER: I don't want you to repeat the lecture, but what were those four points?
LEON: (Long silence) Clyde? (Passage of two or three minutes; seems like an hour.)
LEON: Well, I'd better get back to the office.
CLYDE: Me, too!
WALTER: Me, too!

What can we learn from the Leon-Clyde story?

Both Leon and Clyde were brilliant men, yet neither was able to recall even a fragment of any point made in the previous night's lecture. Each had forgotten

---

[1] Douglas A. Bernstein, Edward J. Roy, Thomas K. Srull, and Christopher D. Wickens, *Psychology* (Boston: Houghton Mifflin, 1988), p. 293.

[2] Alan J. Parkin, *Memory: Phenomena, Experiment and Theory* (Cambridge, MA: Blackwell, 1993); Hermann Ebbinghaus, *Memory* (New York: Dover, 1964), p. 76.

the four points because neither had transferred the points from short-term memory to long-term memory by silently reciting them. Instead, both had recited that the speaker was clear, forceful, and wise and that he had made four points—and they remembered only what they had recited. As you can surmise from the anecdote, the only sure way to overcome forgetting is by taking notes and then studying and reciting them.

What are the elements of successful note taking?

But taking notes doesn't simply mean scribbling down or marking up the things that strike your fancy. It means starting with some sort of system and then gathering information quickly and efficiently before pulling things together.

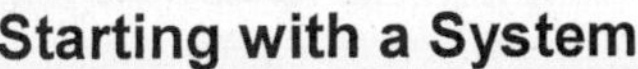

Starting with a System

What can you do to make your notes more meaningful?

Too many students rely on a seat-of-the-pants method of taking notes. Whether you're taking notes for a lecture or for a reading assignment, it pays to have a system in place. If you're taking notes in a lecture or separate notes for a textbook assignment, the Cornell System is your best bet. If you're actually marking up your textbook pages, it's unwise to star, underline, and highlight in a haphazard fashion. You need to follow some consistent guidelines to ensure that your markings are meaningful.

Using the Cornell System

In what sense are notes more valuable than a textbook?

The notes you jot down can become a handwritten textbook. In fact, in many instances they are more practical, meaningful, and up-to-date than a textbook. If you keep your notes neat, complete, and well organized, they will serve you splendidly.

What is the Cornell System?

The best way to ensure that the notes you take will be useful is by adopting the Cornell System, which was developed at Cornell University almost fifty years ago. Over the years, the Cornell System has been embraced by countless colleges and universities in the United States and throughout the world. Although the system is far-reaching, its secret is simple: Wide margins on the outside and the bottom of each page are the key.

What are the principal components of Cornell-style note paper?

Many office and school supply stores now sell Cornell-style note paper, but you can easily use a pen and ruler to adapt standard loose-leaf paper to the task. First draw a vertical line down the left side of each page two-and-one-half inches from the edge of the paper; end the line two inches from the bottom of the sheet. This creates the *cue column.* Next draw a horizontal line two inches up from the bottom of the page. This is the border for your *summary area.* The large space to the right of the cue column and above the summary area is where you write your notes. Figure 3.1 shows a Cornell note sheet.

What are the cue column and summary area used for?

As you're taking notes, the cue column should remain empty, as should the summary area. But when the time comes to review and recite what you've jotted down, you'll use the cue column for questions to help clarify meanings, reveal relationships, establish continuity, and strengthen memory. The summary area will be used to distill a page's worth of notes down to a sentence or two.

How does the Cornell System adjust to the demands of different courses?

The information that goes in the largest space on the page varies from class to class and from student to student. Different courses come with different demands. The format you choose for taking your notes and the ideas you jot down are almost entirely up to you. If you have a special way of jotting down your notes, you should be able to use it with the Cornell note sheet.

Why is the outline format discouraged?

About the only format you should be wary of is the outline. As crisp and neat as they may look at a glance, outlines have a way of tying your hands and forcing you to squeeze information into an unforgiving framework. When you're taking notes—especially in a lecture—the last thing you want to be worrying about is Roman numerals.

What was Professor Fox's objection to outlines?

Edward W. Fox, Cornell University's great teacher, lecturer, and historian, was skeptical about the value of outlines and other rigid setups in note taking:

> Elaborate arrangements tend to confuse, and the traditional topical form, the use of Roman numerals, capital letters, Arabic numerals, and small letters, etc., with much indentation, has a fatal tendency to imply a logical analysis rather than elicit one.[3]

---

[3] Edward W. Fox, *Syllabus for History* (Ithaca, NY: Cornell University Press, 1959). Reprinted by permission of the author.

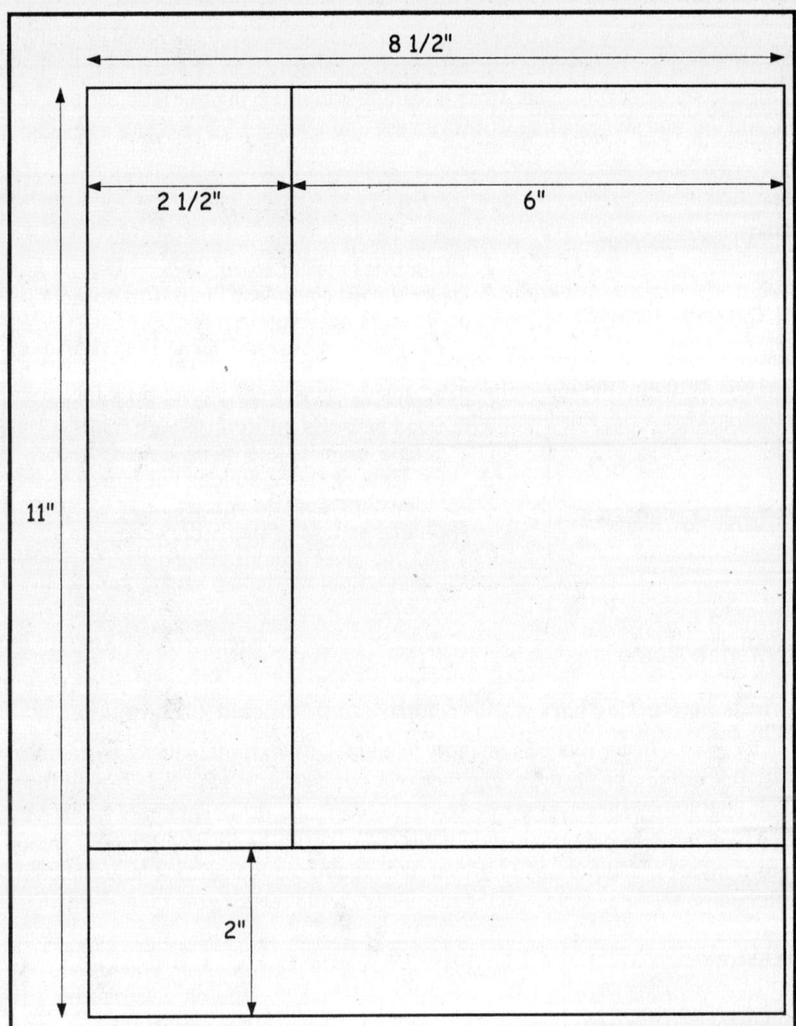

Figure 3.1
The Cornell Note Sheet

What should be your primary focus in all note-taking formats?

In short, you should be concentrating on the information itself, not on its numbering scheme. Here are a few formats that should work well with either lectures or textbook readings.

How do sentences in your notes differ from traditional sentences?

Sentences. Take key ideas from a lecture or reading and jot them down in your own words. If you're taking notes in a lecture, you probably won't have time to write out complete sentences. Instead, write telegraphically, leaving out articles such as *a, an,* and *the,* and abbreviating words you use often. Figure 3.2 provides an example.

	Oct. 10 (Mon.) - Soc. 102 - Prof. Oxford
What's animism?	A. Animism 1. Object has supernatural power 2. Power called mana (not limited to objects)
Describe mana!	a. Objects accumulate mana Ex. Good canoe-more mana than poor one.
How to gain mana?	b. Objects can lose mana c. People collect objects w/lots of mana d. Good person's objects collect mana e. People, animals, plants have mana, too. Ex. Expert canoe builder has mana- imparts mana to canoe
Who has mana?	f. Chief has lots of mana-dangerous to get too close to chief-mana around head.

Figure 3.2
Sentence Notes

How do note-taking paragraphs compare to traditional paragraphs?

Paragraphs. Cluster related sentences in a block of text, often under a heading or label that serves to tie them together. As you see in Figure 3.3, these are not traditional paragraphs where one complete sentence flows smoothly to the next. Your sentences in paragraph-style notes will usually be telegraphic, and smooth transitions are not important. However, these paragraphs share one important thing with traditional paragraphs: All the sentences should relate to the same main idea.

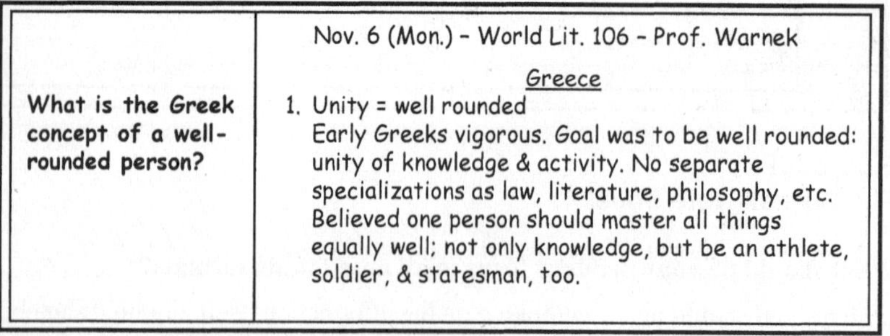

	Nov. 6 (Mon.) - World Lit. 106 - Prof. Warnek
	Greece
What is the Greek concept of a well-rounded person?	1. Unity = well rounded Early Greeks vigorous. Goal was to be well rounded: unity of knowledge & activity. No separate specializations as law, literature, philosophy, etc. Believed one person should master all things equally well; not only knowledge, but be an athlete, soldier, & statesman, too.

Figure 3.3
Paragraph Notes

What do definition-style notes look like?

Definitions. Write a name or term, add a dash or colon, and then provide a succinct explanation or elaboration. Figure 3.4 provides an example.

	Mar. 14 (Fri.) - Ed. 103 - Prof. Pauk
	Types of note-taking formats
What are main types of note-taking formats?	sentence - Notes written in sentences, but telegraphically: w/abbrevs. for common wds. Articles ("a", "an", etc.) left out.
	paragraph - Like real paragraphs, clustered around main idea, but sentences telegraphic & transitions left off
	definition - name or term, followed by dash or colon and explanation
	list - word or phrase heading, followed by series of items. No numbers unless relevant. Use bullet pts. instead.
	combination - mix of other formats

Figure 3.4
Definition Notes

How do you make items stand out in list-style notes?

Lists. Start with a topic, name, term, or process, and then list phrases or telegraphic sentences that relate. Avoid numbering the items unless the numbers are relevant to the list. If you want to make the items stand out in your notes, consider beginning each with a bullet point. Figure 3.5 shows what these lists are apt to look like.

	Mar. 14 (Fri.) - Ed. 103 - Prof. Pauk
	Intonation in textbooks
How do texts show intonation?	* boldface signals heading, subhead. May also indicate key word, phrase
	* italics place emphasis
	* underline like boldface or italics. Depends on format of book
	* bullets (circles, dots, squares, etc.) set off items in lists

Figure 3.5
List Notes

What is the advantage of using drawings in your notes?

Drawings. Drawings and diagrams can succinctly sum up information that may be difficult to explain in words alone. A sketch can often convey locations or relationships more effectively than a sentence or two. Figure 3.6 shows the sort of diagram that a biology student might include in her notes.

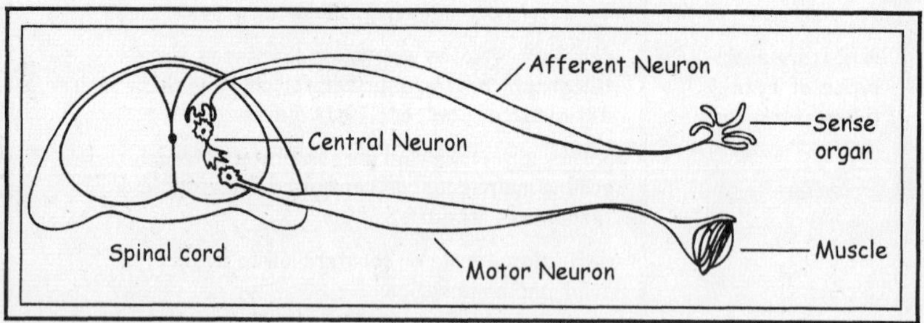

Figure 3.6
Diagramming as a Study Aid

When will you need a combination of formats for your notes?

Combination. Some chapters and a few lectures may fit into a single note-taking format, but it's unlikely that most will. Good note takers must remain flexible, shifting quickly from one format to another to capture key information as efficiently and meaningfully as possible. It's important to choose the right tool for the job. Figure 3.7 shows notes that combine several formats.

Marking Your Textbook

What is your other option when taking textbook notes?

Although the Cornell System can work as well with readings as it does with lectures, when you're taking notes for a textbook assignment, you also have the option of putting those notes directly on the pages of your book instead of jotting them down on Cornell note paper.

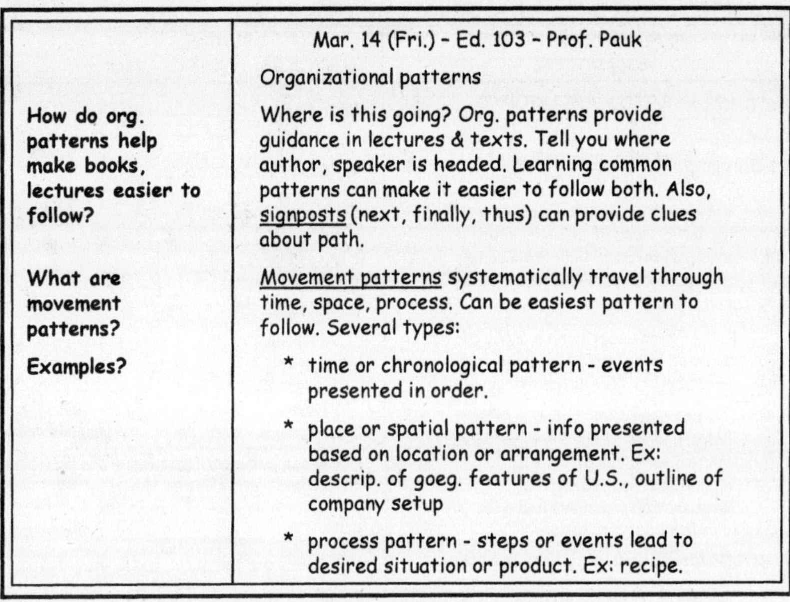

	Mar. 14 (Fri.) - Ed. 103 - Prof. Pauk
	Organizational patterns
How do org. patterns help make books, lectures easier to follow?	Where is this going? Org. patterns provide guidance in lectures & texts. Tell you where author, speaker is headed. Learning common patterns can make it easier to follow both. Also, signposts (next, finally, thus) can provide clues about path.
What are movement patterns?	Movement patterns systematically travel through time, space, process. Can be easiest pattern to follow. Several types:
Examples?	* time or chronological pattern - events presented in order.
	* place or spatial pattern - info presented based on location or arrangement. Ex: descrip. of goeg. features of U.S., outline of company setup
	* process pattern - steps or events lead to desired situation or product. Ex: recipe.

Figure 3.7
Notes That Use a Combination of Formats

What happens to the key Cornell components when you choose to mark your text instead of taking separate notes?

When you choose to take notes directly in your textbook, the outside margin of the textbook page becomes the substitute for the cue column, and the spaces at the bottom and top of each page serve the same purpose as the summary area in your Cornell note paper. The big difference, of course, between a Cornell note sheet and a textbook page is the large note-taking area itself. Because a textbook page is already filled with text, you obviously can't put traditional notes where the text already is. What may at first seem like a disadvantage can actually be a great boon. Rather than jotting down your notes from scratch, you can mark the words or phrases in your textbook that make up the most important ideas.

Figure 3.8 contains twelve suggestions for marking textbooks. Notice especially the use of single and double underlines; the use of asterisks, circling, and boxing for important items; and the use of the top and bottom margins for long notations or summaries. If some of these ideas appeal to you, work them into your marking system. Be sure to use them consistently so that you will instantly remember what they mean.

EXPLANATION AND DESCRIPTION	SYMBOLS, MARKINGS, AND NOTATIONS
1. Use double lines under words or phrases to signify main ideas.	Radiation can produce mutations...
2. Use single lines under words or phrases to signify supporting material.	comes from cosmic rays...
3. Mark small circled numbers near the initial word of an underlined group of words to indicate a series of arguments, facts, ideas—either main or supporting.	Conditions change... (1) rocks rise... (2) some sink... (3) the sea dashes... (4) strong winds...
4. Rather than underlining a group of three or more important lines, use a vertical bracket in the margin.	had known... who gave... the time... of time...
5. Use one asterisk in the margin to indicate ideas of special importance, and two for ideas of unusual importance. Reserve three asterisks for principles and high-level generalizations.	* When a nuclear blast is... ** People quite close to the... *** The main cause of mutations...
6. Circle keywords and key terms.	The (genes) are the...
7. Box words of enumeration and transition.	[fourth,] the lack of supplies... [furthermore,] the shortage...
8. Place a question mark in the margin, opposite lines you do not understand, as a reminder to ask the instructor for clarification.	The lastest... ? cold period... about 1,000,000... Even today...
9. If you disagree with a statement, indicate that in the margin.	Disagree Life became... on land only... 340 million years...
10. Use the top and bottom margins of a page to record ideas of your own that are prompted by what you read.	Why not use carbon dating? Check on reference of fossils found in Tennessee stone quarry.
11. On sheets of paper that are smaller than the pages of the book, write longer thoughts or summaries; then insert them between the pages.	Fossils Plants = 500,000,000 years old Insects = 260,000,000 years old Bees = 100,000,000 years old True Fish = 330,000,000 years old Amphibians = 300,000,000 years old Reptiles = 300,000,000 years old Birds = 150,000,000 years old
12. Even though you have underlined the important ideas and supporting materials, still jot brief cues in the side margins.	Adapt - ______ fossil - ______ layer - ______

Figure 3.8
A System for Marking Your Textbooks

What is the risk of marking up your textbook?

Textbook marking can be a useful aid to study and review, but it must be done sparingly and with some self-discipline. If you overmark your book, you will defeat your purpose: quick identification of important points. When the time comes to master your notes, instead of reviewing ideas you may find yourself trying to decipher a code. Also, bear in mind that the "you" who reviews the marked book will not be quite the same "you" who did the marking. As the term progresses, your knowledge will grow. By the end of the term, you will be accepting as commonplace many things that seemed so important to underscore, box, circle, star, question, comment on, or disagree with at the beginning of the term. Your early marks may hamper your review. So use the help that marking can give you, but don't go overboard. Think ahead.

What advantage does textbook marking have over separate notes?

If you can manage to keep your marking under control, you'll wind up with a valuable study tool. One advantage of marking up your textbook directly is that your notes and your textbook travel together as a unit. Whenever you need clarification for a sentence or phrase you have marked, you have an entire book's worth of supporting material, including a table of contents, an index, and often a glossary, at your fingertips.

What is a potential disadvantage of textbook marking?

Of course, a potential disadvantage of textbook marking is that it doesn't require your complete concentration. When you read a paragraph in your textbook and write down a summarizing sentence on a separate Cornell note sheet, you have already initiated the process of taking the textbook information and making it your own. Contrast this to textbook marking, where it's possible to thoroughly annotate your textbook without truly understanding what you've marked. Drawing lines and boxes and inserting symbols and question marks can give you a false sense of accomplishment if you're not thinking deeply about what you read.

How can you make sure you understand what you've marked?

Luckily, there is a simple but powerful way to solve this problem: recitation. Rather than marking each paragraph as you go, read it through completely and then recite its main idea in your own words. Then and only then should you return to the paragraph and mark the words or phrases that express or support the answer you've just recited. This step will move you out of the realm of simple recognition and into recall. Recalling your notes is the way to help make information your own.

On the following pages are examples of appropriately marked textbook pages. Figure 3.9 shows how to organize a page using enumeration, that is, encircling words such as *first* and *second.* Write in numbers to identify salient points. The underlinings should be sparse and form the answers to the questions in the margins. This type of organization not only aids in comprehending and remembering the main points of the page, but also helps immensely when you're studying for an examination and time is short.

	MARITIME EXPANSION
Ming period	Another dramatic development of the Ming period
Naval expeditions	was the series of naval expeditions sent out
When? Who?	between 1405 and 1433 under Hong Wu's son
	Yong Lu and Yong Lu's successor. China had a
Naval history	strong maritime history stretching back to the
	eleventh century, and these early fifteenth -century
	voyages were a continuation of that tradition. The
Relative power?	Ming expeditions established Ch ina as the greatest
	maritime power in the world—considerably ahead
Portugal power	of Portugal, whose major seafaring reconnaissances
when?	began a half-century later.
Purpose of	In contrast to Honh Wu, Yong Lu broadened
expeditions?	① diplomatic and ② commercial contacts within the
Tribute system??	tribute system. Yong Lu had two basic motives for
2 motives?	launching overseas voyages. First, he sent them in
Contender -	search of Jian Wen, a serious contender for the
who?	throne whom he had defeated but who, rumor
	claimed, had escaped to Southeast Asia. Second, he
	launched the expeditions to explore, to expand
	trade, and to provide the imperial court with
	luxury objects. Led by the Muslim eunuch admiral
Admiral?	Zheng He and navigating by compass, seven fleets
1st expedition	sailed to East and South Asia. The first expedition
	(which carried 27,800 men) involved 62 major
Ship's size?	ships, the largest of which was 440 feet in length
	and 180 feet in the beam and had 9 masts. The
Sea route?	expeditions crossed the Indian Ocean to Ceylon,
	the Persian Gulf, and the east coast of Africa.
3 consequences?	These voyages had important consequences. They
	extended the prestige of the Ming Dynasty
	throughout Asia. ① Trade, in the form of tribute from
	as far as the west coast of southern India, greatly
	increased. ② Diplomatic contacts with the distant
	Middle East led to the arrival in Nanjing of
	embassies from Egypt. ③ The maritime expeditions
	also led to the publication of geographical works.

Figure 3.9
Use of Enumeration in Textbooks

Source: John P. McKay, Bennett D. Hill, and John Buckler, *A History of World Societies.* Copyright © 2000 by Houghton Mifflin Company, Boston. Reprinted by permission.

Gathering Information

What can you do to be fully engaged as you take notes?

Whether you are taking notes on sheets of paper or making marks in your textbook, you need to be fully engaged. To avoid drifting into the sort of robotic routine that adds little to learning, it is important to remain inquisitive as you take your notes, to keep alert for signs that will tip you off to the meaning of what you're reading or learning, to record things efficiently, and, as always, to stay flexible for the inevitable exceptions and special cases.

Being Inquisitive

What is the secret to being an active participant?

You can't really expect to do a good job of gleaning the most important information from a lecture or reading unless you are paying attention. And while getting enough sleep and sitting up straight can help promote alertness, the real secret to being an active participant is to maintain an inquisitive mindset. Formulating questions, whether silently or aloud, will unlock the meaning of information in a way that listening or reading passively just can't approach.

Use Questions to Unlock Meaning

How does Eddie Rickenbacker's story illustrate the effect of questions on understanding?

Why are questions so important? The story of Eddie Rickenbacker provides a dramatic illustration of their effect on understanding. Rickenbacker was a highly decorated World War I flying ace who went on to become chairman of Eastern Airlines. Brought up in humble circumstances, he was forced to leave school at the age of twelve so he could find work and support his widowed mother.

Although Eddie couldn't attend school regularly, he still appreciated the value of an education and later decided to take a correspondence course in mechanical engineering. Here's part of his story:

> The first lesson, I do not mind admitting, nearly finished my correspondence-school education before it began. It was tough. . . . As there was no teacher of whom I could ask a question, I had to work out the answers myself. Once I reached the answer through my own individual reasoning, my understanding was permanent and unforgettable.[4]

---

[4] Edward V. Rickenbacker, *Rickenbacker* (Englewood Cliffs, NJ: Prentice-Hall, 1967), pp. 31–32.

Perhaps you can picture young Eddie Rickenbacker with paper and book spread out on a kitchen table, struggling to gain meaning from a paragraph. Doubts began to mount. Plain grit wasn't enough. He felt overwhelmed. Then, in desperation, he probably said to himself, "All right, Eddie, try it once more: What's this fellow trying to tell me?"

Why did questions provide such a breakthrough in Rickenbacker's learning?

Right then, he created a miniature miracle. How? He asked a question! Armed with this questioning technique, he became his own teacher. Previously, his eyes touched the words on the page and, in touching, expected that meaning would somehow, like a jack-in-the-box, pop up. Unfortunately, it didn't (and it doesn't). But with a question ringing in his ears, Rickenbacker was able to focus on hearing an answer. He heard it and understood it. You can too.

How did Socrates use questions?

We shouldn't be surprised that questions are packed with so much power. Human beings have known this for more than twenty-four hundred years. Socrates (469–399 B.C.), the greatest of the Greek philosophers, developed what is known as the Socratic method, *the questioning method.* By employing a series of carefully directed questions, Socrates would lead another person, through his own step-by-step answers, to arrive at the understanding or conclusion himself.

Why are questions so powerful?

What is it that makes a question so powerful? Maybe, as the psychologists say, questions promote concentration. If true, what a sure and easy way to gain concentration with a technique that is available to all of us!

Ask Questions Both Silently and Out Loud During a Lecture

What are the roles of the speaker and listener in a lecture?

Although the communication may seem one-sided, both the speaker and the listener play important roles in a classroom lecture. The speaker's responsibility is to make points clearly. The listener's responsibility is to understand what the speaker says. If a speaker's message is not clear and the listener asks a clarifying question, both the speaker and the listener benefit. The speaker is encouraged and gratified to know that the audience is interested. The listener can concentrate on what the speaker has to say and feel good about raising a question that other, more timid members of the audience might have been hesitant to ask.

What did the University of Virginia survey reveal about questions?

As vital as questions can be to comprehension, they often go unasked. A professor at the University of Virginia who conducted a survey found that 94 percent of her students had failed to understand something in at least one class lecture during the semester. Seventy percent of the students had not asked clarifying questions even though they knew they could. When she asked them why they had remained silent, they answered with such statements as "I was afraid I'd look stupid," "I didn't want to make myself conspicuous," "I was too proud to ask," and "I was too confused to know what question to ask."

How do you overcome the fear of asking questions?

The way to dispel the fear of asking is to remember that the only dumb question is the one that is never asked. The way to dispel confusion is to acknowledge it by saying, "I'm confused about the last point you made" or "I'm confused about how the example pertains to your main point." In this situation, as in most, honesty is the best policy.

Formulate Questions as You Read

How do you ask questions when you're reading a book that can't talk back?

It's easy to see how questions can help clarify important points in a classroom lecture. After all, you usually have an expert right there who can sense your confusion and respond to the questions. Not so with a textbook assignment, where all you have are silent words on a printed page. Instead of approaching a textbook as a passive recipient of its information, build on the relationship you formed with the author or authors when you first surveyed the book and read over its prefatory material by constantly formulating questions as you read, by wondering out loud about issues or aspects that concern you, and by writing out questions that help you pinpoint and remember the most important information. The latter really serves as the foundation for taking notes and mastering them.

What's a good way to start asking questions as you read?

If asking questions as you read doesn't seem to come naturally at first, a good way to start is by reading the headings and subheadings in your assignment and turning them into questions. This is an important ice-breaker in a number of textbook reading systems, including the well-known SQ3R system (see Figure 3.10).

THE SQ3R SYSTEM	
S SURVEY	Glance through all the headings in the chapter, and read the final summary paragraph (if the chapter has one). This survey should not take more than a minute, and it will show you the three to six core ideas on which the discussion will be based. This orientation will help you organize the ideas as you read them later.
Q QUESTION	Now begin to work. Turn the first heading into a question. This will arouse your curiosity and thereby increase comprehension. It will bring to mind information you already know, thus helping you understand that section more quickly. The question also will make important points stand out from explanatory details. You can turn a heading into a question as you read the heading, but it demands conscious effort on your part.
R_1 READ	Read the paragraph or section to answer the question. Read actively.
R_2 RECITE	After you finish reading the paragraph or section, stop, look away from the book, and try to recite the answer to your formed question. If you cannot recite the answer correctly or fully, reread the section and try again.
R_3 REVIEW	When you have finished reading and reciting page after page, go back to the beginning of the chapter, glance at the headings and subheadings, and think briefly about the answers that you have already recited. Work your way in this manner to the end of the chapter. Now you should have ended with an integrated bird's-eye view of the entire chapter.

Figure 3.10
The SQ3R System

Source: Adaptation of "Steps in the SQ3R Method" (pp. 32–33) from *Effective Study,* 4th Edition, by Francis P. Robinson. Copyright 1941, 1946 by Harper & Row Publishers Inc. Copyright © 1961, 1970 by Francis P. Robinson. Reprinted by permission of the publisher.

It doesn't take much to transform a typical textbook heading or subheading into an attention-getting question. For example, the main heading "Basic Aspects of Memory" could be turned into the question "What are the basic aspects of memory?" The technique is simple, but it works. Here are some additional examples.

Subtopic Heading	Question Formulated
The Memory Trace	What is a memory trace?
Rate of Forgetting	How fast do we forget?
Organization of Recall	How is recall organized?
Decay Theory	What is the decay theory?

What do you do once you've turned a heading into a question?

Once you have turned a heading into a question, you read the material under the heading to answer your question. Suddenly you are reading with a purpose instead of just passively taking in information. If the question is answered early in the discussion, ask another, based on what you have read.

Following the Signs

What signs help improve comprehension of lectures or readings?

Whether information is delivered in the form of a lecture or a textbook chapter, there are usually signs that help direct you down the road to comprehension. Both the intonation of the words and the way they are organized provide clues about the author or speaker's purpose and approach.

Pay Attention to Intonation

How does intonation affect the meaning of words?

Consider the dramatic effect that intonation can have on the meaning of even a simple phrase, such as "Excuse me." Depending on the tone of voice, these two words can sound polite, tentative, argumentative, or downright resentful. In each case, the words are identical, but the delivery is different. If you pay attention only to the words when you're listening to a lecture or reading a textbook, but ignore their delivery, you may be missing a critical component of their meaning.

What are the three components of intonation?

Intonation in Lectures Most college lecturers speak about 120 words per minute, which means that in a fifty-minute lecture you hear roughly 6,000 words. Listening for signals in a lecture is especially helpful because, unlike in reading, you don't have the luxury of retracing your steps if you discover that you're lost. In addition to words, intonation—variations in the lecturer's voice—is the most significant signal in spoken language. Intonation has three components: volume, pauses, and cadence.

1. *Volume.* In general, the introduction of a crucial idea is preceded by a change in volume; the speaker raises or lowers his or her voice.
2. *Pauses.* Pausing before and after main ideas sets these ideas apart from the rest of the lecture. Pauses achieve a dramatic effect and, on a practical level, provide note takers with extra writing time.
3. *Cadence.* The rhythm of a lecturer's speaking patterns can be particularly helpful. Often, like the bulleted lists you find in textbooks, the speaker lists a series of important ideas by using a steady speaking rhythm, sometimes even beginning each idea with the same words or phrase. Whenever you detect these oral signals, your pencil should be moving steadily, adding these important points to your notes.

What does reading with intonation involve?

Intonation in Textbooks Reading with intonation can make the words on your textbook page come alive. This doesn't mean reading out loud, but it does mean reading with expression. Intonation helps you combine individual words into meaningful mental "bites."

How does reading expressively help?

As your eyes move rapidly across the page, let your mind swing along each line with an intonational rhythm that can be heard by your "inner ear." Read the line expressively. In doing so, you will be supplying the important rhythm, stress, emphasis, and pauses that were taken out when the words were turned into written form. This will put the meaning of the words more quickly within your grasp.

How do you develop the habit of reading with intonation?

To make silent intonation a regular habit, take a few minutes to read aloud in the privacy of your room. This will establish your own speech patterns in your mind so you will "hear" them more readily when you read silently.

How do type styles mimic the volume, pauses, and cadence of speech?

Adding your own intonation in this way can help impart meaning and expression to words that may otherwise seem to lack pizzaz. In addition, books provide another form of intonation that mimics the volume, pauses, and cadence of speech.

Open any textbook and you'll quickly discover that the words aren't all printed in the same size or the same style. The format may differ from text to text, but in general each book takes advantage of a variety of type sizes and styles to convey information. By noting these typographical differences, you can pick up on signals for organization and emphasis.

Boldface (thick, dark type) often signals a textbook heading or subheading. It may also be used to draw your attention to a specific principle, definition, or keyword within the text.

Italics (type that slopes to the right) places emphasis on a word or a phrase.

Underlining often performs the same functions as boldface and italics, depending on the format of the particular textbook.

- Bullets (small markers, often circular or square) set off the items in lists.

Size, color, and placement of type often call attention to headings and subheadings. Take note of words printed in larger type, in color, or on lines by themselves.

How can you learn the meaning of a book's typographical intonation?

You can usually crack a book's particular typographical code by skimming through it before you start reading. In addition, look for an explanation of format—especially if it is unconventional—in the book's introductory material.

Recognize Organizational Patterns

What role do organizational patterns play in lectures and textbooks?

Where are we going? And how are we going to get there? Those are both reasonable questions to ask when you're heading off on a journey. They're equally reasonable when you're reading an assignment or listening to a lecture. Luckily, both authors and speakers normally tell you where you're headed and how you're going to get there by using common organizational patterns to help arrange their information. Familiarize yourself with these patterns, and you should find things easier to follow. In addition, you can often navigate through information by noticing certain verbal signposts that commonly line the route. Simple words such as *next, thus,* or *finally* can provide valuable clues to the path that a chapter or lecture is taking.

What are movement patterns?

Movement Patterns Authors or speakers will frequently move you systematically through time, through space, or through a process. Once you catch on, these patterns can be among the easiest to follow.

With the **time or chronological pattern,** events are presented in the order in which they occurred. This pattern can be recognized quickly from the author's or lecturer's use of dates and such phrases as *in previous years, the next day,* and *two years later,* which denote the passage of time.

Items in a **place or spatial pattern** are presented or discussed on the basis of their locations or their arrangement relative to one another. For example, an author might use a spatial pattern to describe the geographical features of the United States from the West Coast to the East Coast. This is often called a *geographical* *pattern.* It is also called a *topical pattern* when it is used to describe the organization of a corporation along the lines of purchasing, manufacturing, sales, and so forth. The progression from item to item is usually orderly and easy to follow: from left to right, from high to low, from north to south, and so on.

Steps or events in a **process pattern** are presented in an orderly sequence that leads to a desired situation or product. A recipe and the instructions for assembling a bicycle are examples of process patterns. They often include words such as *first, after this, then, next,* and *finally.* You'll often encounter this pattern in computer courses and the sciences, where the steps in a process are described in the order in which they must occur to put something together, run an application, or blend ingredients.

How are importance patterns organized?

Importance Patterns A common way of organizing facts or information is by arranging them in terms of their relative importance.

In the **increasing importance pattern,** the most important or most dramatic item in a series is placed at the end. Each succeeding item is more important than the previous one, so a crescendo effect is created. Thus, this pattern is also called the *climactic-order pattern.*

In the **decreasing importance pattern,** the most important or most dramatic item in a series is placed at the very beginning. Such an organization grabs your interest immediately, so there is a good chance that you will stay with the writer or speaker all the way through. This pattern is commonly used in newspaper articles and is known by journalists as the *inverted pyramid pattern.*

What do causal patterns all share in common?

Causal Patterns One thing leads to another in a number of patterns, most of which are variations on the well-known idea of cause and effect.

In the **problem–effect–solution pattern,** the writer or speaker outlines a problem (cause), explains its effect, and then often (though not always) maps out a solution.

The **problem–cause–solution pattern** inverts this approach. A predicament (effect) is introduced, followed by its antecedents (cause) and eventually by remedies. In short, the effect comes first this time, followed by the cause.

Of course, not all cause–effect patterns involve problems. In technical subjects, the generic **cause–effect pattern** is quite common. In this case, variables are defined and their result (effect) is explained. Meanwhile, in the social sciences, you'll often run into the **action–impact pattern,** where some sort of initiative is outlined, such as increased funding for education, followed by a result, such as a higher average income for employees who received a college education.

Regardless of the variation, phrases such as, *as a result* or *consequently,* will usually alert you to a causal pattern.

When do writers and speakers use comparison patterns?

Comparison Patterns Writers and speakers compare things, events, or people when they emphasize similarities, and contrast them when they emphasize differences. Individual characteristics may be compared or contrasted one at a time, or several characteristics may be discussed as a group. In either case, the pattern can be recognized from the various similarities or differences and from the use of words such as *similarly, likewise, conversely,* and *on the other hand.*

What do logical patterns have in common?

Logical Patterns In these patterns, a conclusion is either drawn or stated at the outset and then supported.

With the **inductive pattern,** a number of incidents are cited, then a conclusion is arrived at. The main point will be something like this: "So, on the basis of all these facts, we come to this overriding principle, which is so-and-so."

With the **deductive pattern,** the reverse is true. Here, the principle or general statement is given first, then the events or proofs are enumerated.

What should you do if you can't identify a precise organizational pattern?

Signposts Although identifying the precise organizational pattern can be extremely helpful, it isn't always easy or even possible. Some textbook chapters may defy easy organization, and some lectures can be rambling. In these cases, all is not lost. Keep in mind that your key goal while taking notes is to use cues and patterns to help arrange your thoughts and aid your search for meaning. Luckily, the signposts that often tip you off to a particular organizational pattern can still be helpful even if you have a tough time nailing down the overall arrangement. Table 3.1 lists some common signposts and the directions they're likely to point you in.

Recording Efficiently

What are the key factors in taking notes efficiently?

Depending on whether you're taking lecture notes or textbook notes, you may find that you are short on space, short on time, or both. There are ways that you can take notes more efficiently to avoid any and all of these problems.

How do you make your notes both legible and speedy?

Strive to make your note taking both speedy and sparing. Of course, if you scribble down information too quickly, your notes may be illegible. And if you're too choosy about what you record, you could be left with costly gaps in your information. The way to circumvent these problems and record legible, useful notes at a reasonable speed is to adopt the modified printing style, use telegraphic sentences, and record selectively.

Table 3.1
Signposts

Categories and Examples	*When you come across these words, immediately think . . .*
Example Words	
specifically to illustrate for example for instance that is	"Here comes an example. Must be double-checking to make sure I understood the point just made."
Cause-and-Effect Words	
consequently therefore as a result if . . . then accordingly thus, so hence	"There's an effect word. Better check back when I have a chance to make sure I can find the cause now that I know what the effect is."
Enumeration Words	
the four steps . . . first, second, third next, finally	"That's a lot of steps. I'd better be sure I'm keeping track of all of them and getting them in the right order."
Addition Words	
furthermore as well as along with in addition moreover also not only . . . but also	"Seems there's always something else to be added. Must be worth remembering."
Contrast Words	
on the other hand in contrast conversely although however, despite whereas	"Here comes the other side of the coin. Let's see how it differs from what's been said already."
Comparison Words	
likewise similarly comparatively identical	"Lots of similar things, it seems."
Swivel Words	
however nevertheless yet but still	"Looks like there's going to be a little bit of doubt or 'give back' on the point just made. Better pay attention to this qualifying remark."

Table 3.1
Signposts—*continued*

Categories and Examples	*When you come across these words, immediately think . . .*
Concession Words to be sure indeed though, although granted of course	"Okay! Here comes an argument or two from the opposing point of view."
Emphasis Words more important above all remember in other words finally	"Looks as though what's coming up is going to be important."
Repeat Words in other words it simply means that is briefly in essence as we've seen	"Here comes another explanation. Maybe I'll understand this one a little better."
Time Words before, after formerly, soon subsequently prior, during meanwhile	"Hmm! A time relationship is being established. Let's see: What came first, what came last, and what came in-between?"
Place Words above below beyond adjacent to	"Okay! I'll put these ideas and facts not only in their proper places, but also in their proper relationship."
Summary Words for these reasons on the whole in conclusion in a nutshell to sum up in short finally	"Good. Now I'll get a simple wrap-up of the points that have been made. It's almost sure to be full of key ideas."
Test Words (lectures) This is important. Remember this. You'll see this again. Here's a pitfall.	"Sounds like a potential test item. Better be sure to pay close attention to it."

Use Modified Printing

What is modified printing?

Poor handwriting need not keep you from taking legible notes. The *modified printing style* combines the rapidity of writing with the legibility of printing. Letters are formed smoothly, as with cursive or longhand writing, but are punctuated with the sort of stops and starts characteristic of printing. Your words take on a cursive look, and the periodic breaks between letters prevent your writing from eroding into an unreadable blur.

Why is modified printing so effective?

What makes the modified printing style so effective and easy to learn is that it combines your style of printing with your style of cursive in a mixture that brings out the best elements of both. Here's how:

a b c d e f g h i j k l m n o p q r s t u v w x y z

Figure 3.11 shows how modified printing looks in a typical paragraph.

> *There are four advantages to using this modified printing style. First, it is faster than cursive writing; second, it is neater, permitting easy and direct comprehension; third, it saves time by percluding rewriting or typing; and fourth, it permits easy and clear re-forming of letters that are ill-formed due to haste.*

Figure 3.11
Modified Printing Style

Take Notes Telegraphically

What does it mean to take notes telegraphically?

The best way to take notes is telegraphically. Long before e-mail and the fax machine were invented, important business and personal messages were sent by telegraph. The sender paid by the word; the fewer the words, the lower the cost. A three-word message such as "Arriving three pm" was much less expensive than an eleven-word message: "I will arrive home promptly at three o'clock in the afternoon." Of course, taking notes doesn't cost money, but it does cost time.

How do you take telegraphic notes?

To save valuable time when you're taking notes, leave out unnecessary words such as articles (*a, an, the*), abbreviate words you use often (see Figures 3.12 and 3.13), and streamline definitions by using a colon (:) or a dash (—). Two examples of this telegraphic style are shown in Figure 3.14.

Symbol	Meaning	Symbol	Meaning
+	plus, positive, and	↕	vibration, motion
-	minus, negative	log	common logarithm
×	algebraic *x*, or multiplied by	ln	natural logarithm
÷	divided by	∈	base of natural logarithms
≠	does not equal	π	pi
≈	equals approximately, approximates	∠	angle
>	greater than, greatly, increased, increasing	⊥	perpendicular to
		||	parallel to
<	less than, reduced, decreasing	a°	*a* degrees (angle)
~	sine curve, cosine curve	*d*	*a* minutes (angle)
→	approaches as a limit, approaches	*d'*	*a* seconds (angle)
≥	greater than or equal to	∫	integral, integral of, integration
≤	less than or equal to	*f*	frequency
≡	identical to	f_n	natural frequency
μ	varies directly as	*cps*	cycles per second
∴	therefore	*m*	mass
$( )^{1/2}$	square root	Φ	phase
$( )^{n}$	nth root	*F*	force
vs	versus, against	/	ratio, the ratio of
⏚	ground	⊤⊤⊤	base, support, mount, foundation
↔	varied, variation		
□	area	(	curve, curvilinear

Figure 3.12
Examples of Technical Symbols

anlys	analysis	*pltg*	plotting
ampltd	amplitude	*reman*	remain
asmg	assuming	*rsnc*	resonance
cald	called	*rltnshp*	relationship
cnst	constant	*smpl*	simple
dmpg	damping	*smpfd*	simplified
dmnsls	dimensionless	*stfns*	stiffness
dfln	deflection	*systm*	system
dfnd	defined	*sgnft*	significant
dstrbg	disturbing	*ths*	this
eftvns	effectiveness	*trnsmsblty*	transmissibility
frdm	freedom	*thrtly*	theoretically
frcg	forcing	*valu*	value
gvs	gives	*wth*	with
hrmc	harmonic	*whn*	when
isltr	isolator	*xprsd*	expressed
isltn	isolation		

Figure 3.13
Typical Technical Abbreviations

Reprinted by permission from G. H. Logan, "Speed Notes for Engineers," *Product Engineering,* September 30, 1963. Copyright © 1963 by Morgan-Grampian, Inc.

Take Notes Selectively

How do you take notes selectively?

Of course, you can't and shouldn't jot down every word. With textbook note taking it's impractical. In lectures it's an impossibility. Taking thorough notes, regardless of the format you choose, should not mean writing down everything you hear. Your emphasis should be on the ideas, not the words. And you don't want all the ideas, either, just the key ones (as Figure 3.15 shows), along with any details or examples you need to make those ideas easier to understand.

Dealing with Special Cases

What should you do in special cases?

Not all note-taking scenarios are ideal. If you can't make it to a lecture or have difficulty taking notes when you do, you need an alternate plan. In the same way, there are factors that can throw a wrench into traditional textbook note taking.

Lecture's words

In marketing, we try to understand customers' needs and then respond to them with the right products and services. In the past, firms often produced goods first and tried to fit the customer's needs to the goods. Today's world-class marketers pride themselves on their customer orientation. We begin with the customer and build the product or service from there. A good example is McDonald's, the fast-food chain, which tailors its menus to local tastes and customs when it opens fast-food outlets in Moscow and other international locations.

Student's telegraphic sentences

Marketing understands customers' needs first.

- In past, firms produced goods first, then fit them to customers.
- World-class = having customer orientation.
- Ex. McDonald's in Moscow

Lecture's words

The US Patent Office has granted numerous patents for perpetual motion machines based upon applications with complete detailed drawings. Some years ago, though, the patent office began requiring working models of such a machine before a patent would be granted. Result: No patents granted for perpetual motion machines since that time.

Student's telegraphic sentences

Perpetual motion machine (drawings) = many patents.

Required working model = no patents since.

Figure 3.14
Examples of Telegraphic Sentences

	Oct. 10 (Mon.) - Soc. 102 - Prof. Oxford
	A. Two kinds of magic
What's sympathetic magic?	1. Sympathetic — make model or form of a person from clay, etc. — then stick pins into object to hurt symbolized person.
Describe contagious magic.	2. Contagious magic a. Need to possess an article belonging to another person. b. Ex. Fingernail clippings. By doing harm to these objects, feel that harm can be transmitted.

Figure 3.15
Selective Note Taking

Have a Backup Plan When You Can't Attend the Lecture

What's the best strategy if you have to miss a lecture?

If you know you'll be missing a class, supply a friend with a cassette or tape recorder and ask him or her to tape the lecture for you. Then you'll be able to take your own notes when you play the tape back.

Use the Two-Page System for Fast-Moving Lectures

How can you cope with a lecturer who speaks too rapidly?

When you need to scramble to keep up with a fast-talking lecturer, you may find this two-page system helpful. Here's how it works: Lay your binder or notebook flat on the desk. On the left-hand page, record main ideas only. This is your primary page. On the right-hand page, record as many details as you have time for. Place the details opposite the main ideas they support. After the lecture, remain in your seat for a few minutes and fill in any gaps in your notes while the lecture is still relatively fresh in your mind.

Take a Different Approach for Supplemental Readings

Why do instructors assign outside reading?

In many undergraduate courses, assignments and lectures focus on a single textbook, but instructors often assign outside reading in other publications. Reasons for assigning the extra work include the following:

1. To amplify topics treated in the textbook or mentioned in class lectures
2. To go into greater detail—for example, by assigning original documents or primary sources
3. To expose students to other points of view or different philosophies
4. To bring background material into discussions

What's the best approach to supplemental readings?

Instructors generally do not expect you to master such supplementary material as thoroughly as you master your textbook. Nevertheless, once the assignment has been made, you must cope with it. Clearly, you cannot spend an inordinate amount of time, but you must learn something from your supplemental reading. Here are some suggestions for doing so:

1. Try to figure out why the book was assigned. You might ask the instructor. If you find out, then you can skim the book, looking for pertinent material and disregarding the rest.
2. Read the preface. As you already know from the previous chapter, the preface provides inside information. It may tell you how this book is different from your textbook.
3. Study the table of contents. Notice especially the chapter titles to see whether they are like those in your textbook. If the chapters with similar titles contain the same information as the chapters in your textbook, read the chapters that do not duplicate your textbook's coverage. (Do this with topics covered in classroom lectures, too.)
4. If you have not yet found an "angle," read the summarizing paragraph at the end of each chapter. Make brief notes on each chapter from the information thus gained. With these notes spread out before you, try to see the overall pattern. From the overall pattern, come up with the author's central thesis, principle, problem, or solution.
5. Don't put the book away with only a vague notion of what it is about. You must come up with something so definite that you can talk about it the next day or write about it two weeks later. Do not waste time on details, but be ready to answer general questions: What was the author's central approach? How was it different from that of your textbook? How was it the same? Look for the central issues around which everything else is organized.
6. Have the courage to think big. If you lack courage, you'll waste time on minor details that you won't remember. Select the big issues and concentrate on them.

When a highly condensed summary of a book or of a long selection is required, you need a special approach. The introduction–thesis–body–conclusion sequence is useful in forcing you to understand the material and the way the author develops and supports it. Furthermore, a summary that follows this sequence can be highly condensed; you may be able to capture the main ideas of a collateral book in only a page or two of notes. Figure 3.16 is an example.

	I. Introduction
Experiment in living close to nature.	Thoreau voluntarily withdrew from civilization which he felt was getting too complicated. He spent 2 yrs., 2 mos., and 2 days living at Walden Pond to regain the simplicity of life that comes when one lives close to the soil.
	II. Thesis
Each man and woman should pause to decide just how they should spend their lives. Are they paying too dearly for unessentials?	In a complex civilization, the fast flowing current of unessentials stemming from custom, tradition, advertising, etc., somehow sweeps a person away from the genuine goals in life. Only by temporarily cutting oneself off from civilization, could people realize that their lives need not be so complex. By getting back to nature to rethink the basic issues of life people can chart their course, and attempt to steer their lives in accordance with these standards (not expediences set up by the pressure of complex civilization.)
	III. Body
People should awaken and become aware of real life.	Thoreau did not wish to hold up progress or civilization; rather, he wished that people would be more contemplative and selective in their actions. Thoreau chronicled his experiences at Walden Pond. He wanted to become familiar with nature. a. He built his own hut. b. Average cost of living a week was 27 cents. c. He observed nature: trees, birds, animals, etc.
Live simply & you will live more fully.	He believed that every person ought to measure up to the best they could do. What the best is, depends upon the individual. To have a standard to measure up does not mean that all must have the same, but every one should measure up to a standard in the best way she or he is able to.
	IV. Summary
Urged people to reject unessentials, and get back to fundamentals.	Thoreau wanted to demonstrate that many so-called necessities were not necessary at all. He wanted people to observe, appreciate, and evaluate what was important in life. Once people set their sights upon the good life, they should follow their sights without compromising.

Figure 3.16
Notes in the Form of a Highly Condensed Summary

Pulling Things Together

Why is it a mistake to end your note taking abruptly?

As a lecture draws to a close or a chapter comes to an end, there's a great temptation to just pack up and move on. That may be understandable, but it can prove to be a great waste of an especially valuable time.

Leaving Lectures

What should you do at the end of a lecture?

The closing minutes can sometimes prove to be the most important part of an entire lecture. Speakers who do not pace themselves well may have to cram half the lecture into the last five or ten minutes. Record such packed finales as rapidly as you can. After class, stay in your seat for a few extra minutes to write down as much as you can remember.

What should you do once you've left the lecture room?

As soon as you leave the lecture room, while walking to your next class, mentally recall the lecture from beginning to end. Visualize the classroom, the lecturer, and any chalkboard work. After mentally recalling the lecture, ask yourself some questions: What was the lecturer getting at? What was the central point? What did I learn? How does what I learned fit with what I already know? If you discover anything you don't quite understand, no matter how small, make a note of it and ask the instructor to explain it before the next class.

Finishing Readings

How do you conduct a quick overview of a reading assignment?

After you've completed a reading assignment, step back and quickly overview what you've just read. Here are two ways to do so:

1. *Reread the abstract, introduction, or summary.* Any of these three common elements provides a brief overview of what you've just read and puts the ideas you've picked up in an appropriate context.
2. *Reread the title and headings.* If the text doesn't include an obvious overview, create one yourself by rereading the title, headings, and subheadings. In combination, these elements can help you mentally organize the information you've just learned. Don't spend a great deal of time doing this rereading. The primary purpose is to refresh your memory of the important points so

that you'll be able to focus more carefully on them later. If you find you have questions, jot them down so you can ask them in class.

Summing Up

What are the basic steps in taking lecture and textbook notes?

Despite some obvious differences, the core concepts that underlie taking lecture notes, taking textbook notes, and marking up your textbook are remarkably similar. If you are careful to devise a system and stick with it, to gather information actively, efficiently, and flexibly, and to pull things together conscientiously at the end of a note-taking session, the handful of steps below should put you in a powerful position to master your notes and make them your own.

How to Take Lectures Notes

Figure 3.17 is a flow chart of this process.

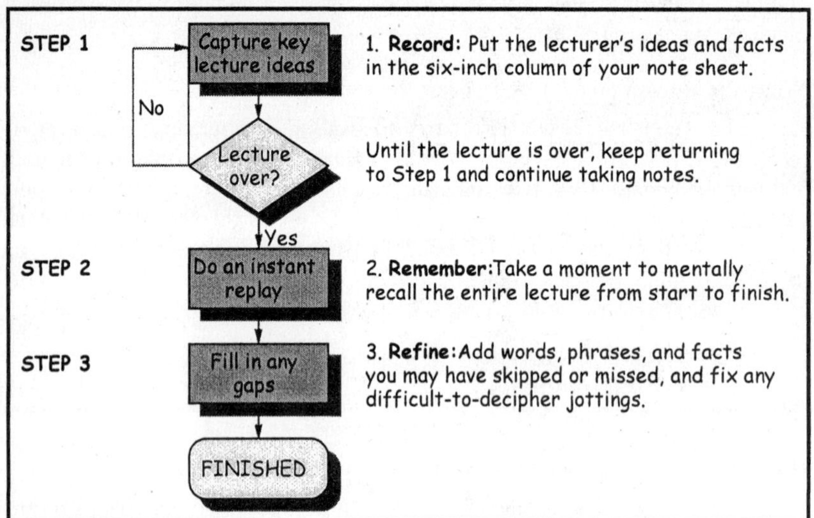

Figure 3.17
Taking Lecture Notes

1. **Record.** Put the lecturer's ideas and facts (along with any relevant diagrams) in the six-inch column of your Cornell System note sheet.
2. **Remember.** As soon as class is over, take a moment to mentally recall the entire lecture from start to finish.
3. **Refine.** Looking over your note sheets, add words, phrases, and facts you may have skipped or missed, and fix any difficult-to-decipher jottings.

How to Take Textbook Notes

Figure 3.18 shows how this process works for textbook marking and separate notes.

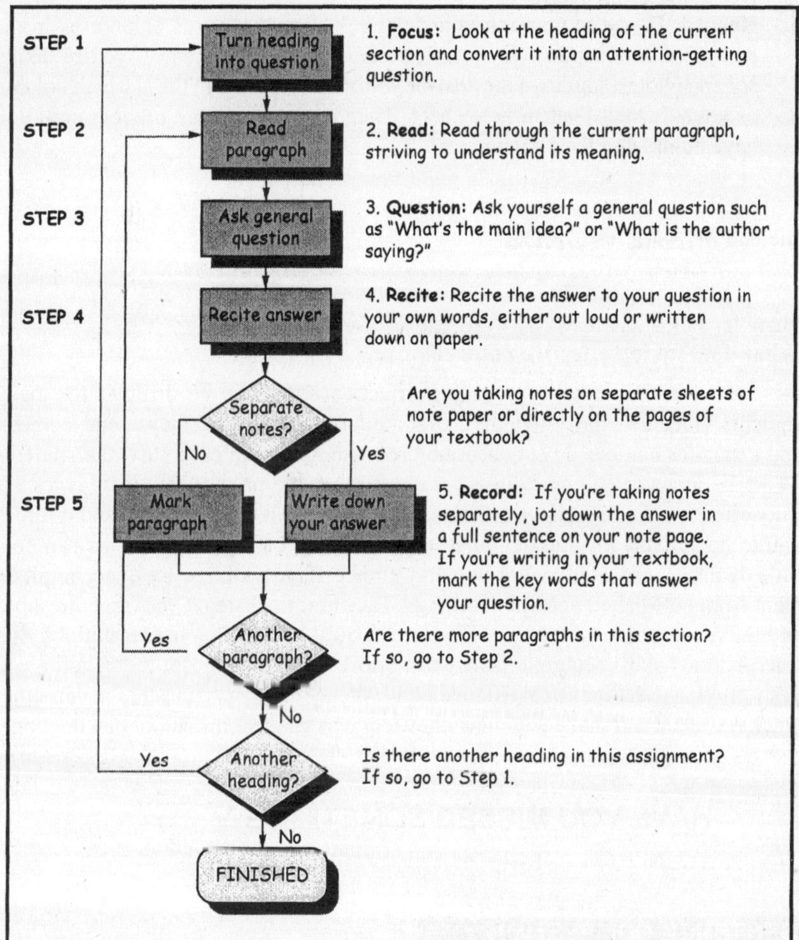

Figure 3.18
Taking Textbook Notes

1. **Focus.** As you read through each section, turn the headings and subheadings into attention-getting questions that help stimulate your thinking and your curiosity as you read on.
2. **Read.** Move through the assignment a paragraph at a time, striving to understand the meaning of the current paragraph.

3. **Question.** Before you move on to the next paragraph, take a moment to ask yourself, "What's the main idea?" or "What is the author saying?"
4. **Recite.** Recite the answer to the question you just asked, in your own words and preferably out loud.
5. **Record.** If you're marking your textbook, find the words or phrases in the paragraph that support your recited answer and mark them. If you're taking separate notes, jot down the answer you've just recited in the six-inch column of your Cornell System note sheet. Then repeat the entire process until you have completed the assignment.

FINAL WORDS

What does taking effective notes require?

Despite indisputable benefits, taking effective notes can be difficult for many students. Of course, most students take some sort of notes when they attend lectures; and a smaller number do so when they're reading. But in each case, the students who take notes often come close but don't reach the point where their notes are truly effective. Taking effective notes requires work; it requires time; and it forces you to be actively engaged in what you're reading or listening to. This can be a little daunting, especially when you may already have a false sense of accomplishment from half-baked notes or markings. Take that final step. Cross that threshold and make your note taking truly effective. You'll be working harder initially, it's true. All new skills require a little extra effort at the outset. But the benefits you derive will materialize almost immediately when you begin to master the notes that you have taken and make your new knowledge permanent instead of just fleeting.

HAVE YOU MISSED SOMETHING?

SENTENCE COMPLETION

Complete the following sentences with one of the three words listed below each sentence.

1. The Leon-Clyde story illustrates the rapidity and scope of ______________.

   note taking **concentration** **forgetting**

2. Words such as *next, thus,* and *finally* often function as ______________.

   signposts **diversions** **intonation**

3. The margin area at the outside of each page should be reserved for ______________.

   reminders **questions** **definitions**

MATCHING

In each blank space in the left column, write the letter preceding the phrase in the right column that matches the left item best.

_____ 1.	Socratic	a.	Time-tested system for recording and mastering your notes
_____ 2.	Drawings	b.	Ancient method that uses questions and answers to promote understanding
_____ 3.	Summary area	c.	Provide clues for the organization of a chapter or lecture
_____ 4.	Telegraphic	d.	Used for distilling a page's worth of notes down to a few sentences
_____ 5.	Two-page	e.	The only note-taking format that is explicitly discouraged
_____ 6.	Outline	f.	Can succinctly describe ideas that are difficult to put in words
_____ 7.	Signposts	g.	System for taking notes in especially fast-paced lectures
_____ 8.	Cornell	h.	Note-taking style that employs only the most essential words

TRUE-FALSE

Write T *beside the* true *statements and* F *beside the* false *statements.*

_____ 1. Memory alone is sufficient for holding on to key ideas from textbooks and lectures.

_____ 2. It's unlikely that a chapter or lecture will fit into a single note-taking format.

_____ 3. A sketch can rarely convey locations or relationships as clearly as a few sentences.

_____ 4. The "you" who took your notes may not be the same "you" who reviews them.

_____ 5. You're seldom expected to read supplemental materials as thoroughly as your textbook.

MULTIPLE CHOICE

Choose the word or phrase that completes each sentence most accurately, and circle the letter that precedes it.

1. To take notes effectively, you need
   a. a loose-leaf notebook.
   b. a system.
   c. sheets of unlined paper.
   d. several fundamental questions.

2. As you're taking notes, the cue column should
   a. remind you.
   b. get wider.
   c. remain empty.
   d. be filled.

3. One advantage of taking notes directly in your textbook is that
   a. your book is more valuable when you sell it at the end of the semester.
   b. your notes and your textbook can travel together as a unit.
   c. you can use a highlighter to single out important ideas and concepts.
   d. as the term progresses, your knowledge is likely to grow.

4. The story of Eddie Rickenbacker underscores the importance of
   a. airlines.
   b. correspondence courses.
   c. questions.
   d. textbook note taking.

5. Modified printing is
   a. speedy, like writing.
   b. neat, like printing.
   c. easy to learn.
   d. all of the above.

SHORT ANSWER

Supply a brief answer for each of the following items.

1. Discuss the principal components of the Cornell System.
2. Explain the various formats for taking notes.
3. Explain how organizational patterns can help increase understanding of lectures and readings.
4. Contrast the process of taking lecture notes with that of taking textbook notes.

WORDS IN CONTEXT

From the three choices beside each numbered item, select the one that most nearly expresses the meaning of the italicized word in the quote. Make a light check mark (✓) next to your choice.

No amount of *sophistication* is going to *allay* the fact that all your knowledge is about the past and all your decisions are about the future.

—Ian E. Wilson (1941–), chairman, General Electric Corporation

1. *sophistication*	argument	refinement	discussion
2. *allay* the fact	change	resolve	relieve

There is always an easy solution to every human problem—neat, *plausible* and wrong.

—H. L. Mencken (1880–1956), American editor and critic

3. *plausible* reasonable smart advantageous

A committee is a *cul-de-sac* down which ideas are lured and then quietly strangled.

—Sir Barnett Cocks (1907–), English scientist

4. a *cul-de-sac* trap net dead end

THE WORD HISTORY SYSTEM

Anecdote
unpublished notes

anecdote an'-ec-dote' *n.* 1. A short account of an interesting or humorous incident. 2. Secret or hitherto undivulged particulars of history or biography.

Even among the ancient Greeks there were two kinds of stories—those given out publicly and those known only privately. The latter kind was called *anekdotos,* "not published." The word was formed by combining *a, an,* "not," and *ekdotos,* "given out." From this source comes French *anecdote* and thence English *anecdote,* which originally retained the Greek significance of "unpublished narratives." But an "unpublished narrative" especially about interesting things and famous people, has a ready market; so *anecdotes* are eagerly brought out on every occasion, and the word loses its original sense, coming to mean simply "a story," "an incident."

Reprinted by permission. From *Picturesque Word Origins* © 1933 by G. & C. Merriam Co. (now Merriam-Webster, Incorporated).

PART

III

Test-Taking Skills

4 Answering Objective Tests

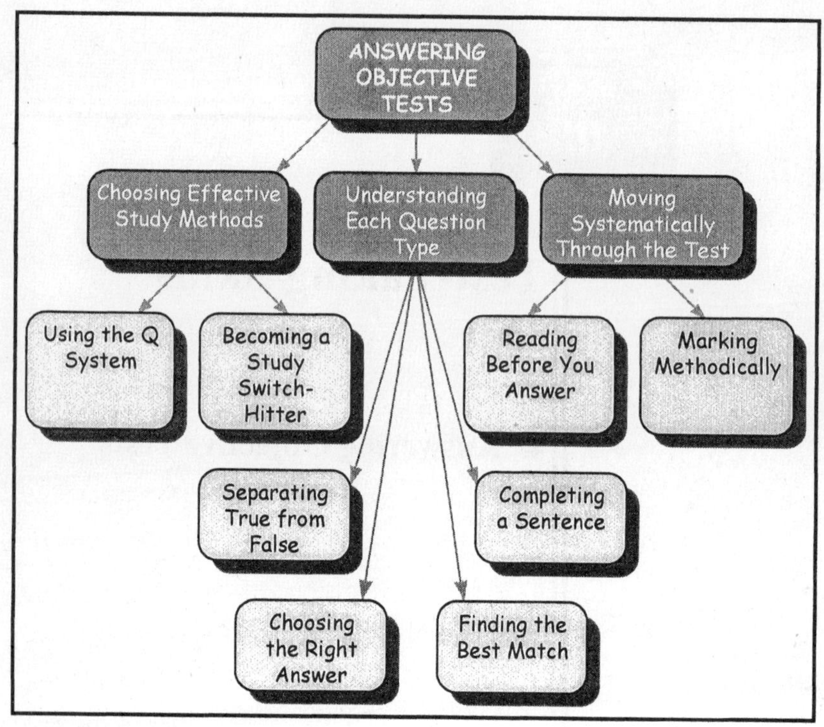
ANSWERING OBJECTIVE TESTS
Choosing Effective Study Methods
Understanding Each Question Type
Moving Systematically Through the Test
Using the Q System
Becoming a Study Switch-Hitter
Reading Before You Answer
Marking Methodically
Separating True from False
Completing a Sentence
Choosing the Right Answer
Finding the Best Match

CHAPTER

4

Answering Objective Tests

What is the answer? . . . In that case, what is the question?

Gertrude Stein
(1874–1946),
American writer;
reportedly her last words

The real purpose of objective tests is to test your knowledge, not to try your patience. Yet when faced with the prospect of answering true-false, multiple-choice, matching, or sentence-completion questions, many students would prefer to choose "none of the above." What they may not realize is that becoming an objective-test expert is easy. To show you how, this chapter looks at:

- **Choosing Effective Study Methods**
- **Understanding Each Question Type**
- **Moving Systematically Through the Test**

What are the two basic categories of test questions?

Most of the test questions you'll be expected to answer in college fall into one of two categories: essay questions and objective questions. Essay questions take a broader view of a subject and generally emphasize your ability to recall and organize what you've learned and to write about it. Objective questions focus more on details and on your ability to recognize, rather than recall, them.

What does mastering an objective test involve?

Mastering an objective test involves choosing an effective study method for committing material to memory, understanding each question type, and moving systematically through the test.

Choosing Effective Study Methods

What are the best study methods to prepare you for an objective test?

As we've already discovered, your success on a test is directly related to how effectively you've studied for that test. You must master your material as efficiently as possible but in a way that will prepare you for any type of test question. The best way to prepare for an objective test is to use the Q System and to be able to recall your material from multiple directions instead of just one.

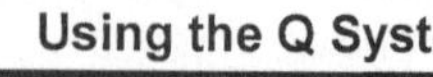

Using the Q System

How can you tackle an objective question with confidence?

The safest way of preparing yourself for an objective test is the safest way of preparing yourself for any kind of test—by studying to the point of recall. Although objective exams and quizzes generally test your ability to recognize, rather than recall, information, learning your notes to the point of recall gives you far greater control over what you've learned. You can tackle a question with confidence when you arrive at the answer independently of the cues that the rest of the question offers.

Why is the Q System such a logical way to study for objective tests?

For that reason, the best way of studying for an objective test is by mastering your notes with the Q System (see Chapter 11 in *How to Study in College* for more details). This system of adding a question alongside each important fact or idea provides your notes with a built-in method for mastery. With the Q System, each question becomes a cue that compels you to recall the idea it refers to. You'd be hard-pressed to find a more logical way of preparing for a test filled with questions than by mastering your notes using questions as cues.

Becoming a Study "Switch Hitter"

What is a study "switch hitter"?

Baseball players who bat from both sides of home plate are more flexible than those who can hit only right-handed or left-handed. In the same way, you can often improve your test-taking average when you master the material in your notes from both directions instead of from just one. Use the Q System as you normally would to recall each important idea, but reverse the process from time to time by covering your questions, reading your notes, and then seeing whether you can remember the questions you wrote to accompany each important idea.

What helps you master your material regardless of its order?

If you have time and want to make absolutely certain you know your material, write each important idea from your notes on the front of a separate index card, and jot your cue on the back. A stack of cards, instead of a few sheets of paper, enables you to constantly rearrange your notes, ensuring that you will be able to recall important information regardless of the order in which it's presented.

Understanding Each Question Type

What is the "trick" to taking objective tests?

In the strictest sense, there are no tricks for taking objective tests. The requirement for taking any test is basically the same: Know your material. Even so, it helps to be acquainted with the quirks and characteristics of each basic question type so you can feel equally at ease whether your test consists of true-false, multiple-choice, matching, sentence-completion, or a mix of all four types of questions.

Separating True from False

What makes a true-false question's simple premise more difficult?

The basic idea behind a true-false question is simple: Faced with a single statement; your job is to decide whether it's true. What makes the choice more difficult is that to be true, the statement must be 100 percent true, not 50 percent or even 99 percent true. One word can be all it takes to turn a true statement into a false one. Consider, for example, the impact of a word like *always* on a true-false statement or how words like *no* and *not* can radically change a statement's meaning. You have to be especially careful about reading each statement thoroughly before you answer it. Look at the following example:

T F In 1787, the year the United States ratified the Constitution, Washington, D.C., became our nation's capital.

The statement is false. Although the Constitution was ratified in 1787 and the nation's capital is Washington, D.C., the United States had no federal capital until 1790, when Congress chose Philadelphia. Washington didn't officially become the capital until June 10, 1800.

What are your chances of guessing correctly in a true-false test?

Although you have a fifty-fifty chance of answering a single true-false statement correctly, the odds are not that high for the entire test. In fact, your chances of guessing correctly on every statement decrease geometrically with every question. In a ten-question test, the odds on guessing are against you by more than a thousand to one. If you're unsure whether to mark "True" or "False," and you're forced to guess, adopt the following two strategies to influence your decision and improve your odds.

Mark "True" If You're Stumped

Why should you mark "True" as your answer if you are stumped?

Because instructors would rather leave true information in your mind, they tend to stack true-false tests with more true statements than false ones. You shouldn't guess right away on a true-false question, but if you're stumped and pressed for time, the odds are in your favor if you choose true over false.

Be Suspicious of Longer Statements

Why are longer statements more likely to be false?

Remember the importance of context, and remember that true-false statements must be 100 percent true. Each word added to a true-false statement increases its chances of being false. All it takes is one incorrect word to make the statement false.

Choosing the Right Answer

What is the basic setup for a multiple-choice question?

A multiple-choice question normally begins with an incomplete sentence known as a stem, followed by a series of choices, known as options, for completing that sentence. In most cases, your job is to find the option that best completes the stem. Here is an example:

In 1787, the year the United States ratified the Constitution,

a. George Washington became the country's first president.
b. Washington, D.C., became the nation's capital.
c. New Mexico was admitted to the Union.
d. the country had no official capital.

In this question, connecting option (d) to the stem results in a true statement. Linking the stem to any other option results in a false statement.

Answering multiple-choice questions entails problems you won't encounter when you're taking a true-false test:

How can directions complicate things in a multiple-choice test?

Varying directions. Some multiple-choice directions tell you to pick more than one correct option; others ask you to mark the one option that is incorrect. Be sure to read the directions carefully and go over all the options before you mark your selection.

What does divided context do to make multiple-choice harder?

Divided context. Because each choice in a multiple-choice question is usually divided into stem and option, you have to mentally connect the two components to determine whether an option is correct. Correct answers aren't always obvious, even when you know your material.

What is the effect of a differing format in multiple-choice questions?

Differing format. Most multiple-choice questions follow the incomplete stem and option format. In some cases, however, the stem may be made up of an entire sentence. A setup of this sort can take you by surprise if you're expecting a standard multiple-choice question, but in general this variation is easier because you don't have to work with a divided context.

You can use several more strategies to cope with multiple-choice questions.

Pick "All of the Above" If You're in Doubt

Why is "all of the above" a good guess?

Most multiple-choice questions present just a single fact, the option that correctly completes the stem. But with "all of the above," the test maker can include several options instead of just one. As the purpose behind a quiz or exam is not only to test but also to teach, "all of the above" becomes an attractive choice for the test maker.

Here's an example of a question that uses "all of the above":

Until the first half of the second millennium B.C., an army laying siege to a city could use

a. scaling ladders.
b. siege towers.
c. archery fire.
d. all of the above.

The correct answer is (d).

What's one way to confirm that "all of the above" is correct?

One way to confirm the choice "all of the above" is to pick out two correct answers in the options. For instance, in the last example, suppose you are sure that ladders and towers were used, but aren't certain about archery fire. Unless the directions permit you to mark more than one option, you already have all the information you need to choose the correct answer. If option (a) is correct and option (b) is correct, option (d) is the only logical answer.

Should you always choose "all of the above"?

It would be a mistake to mark every "all of the above" you run into before reading the question and carefully considering the options. But if you can't seem to come up with an answer and you're running out of time, choosing "all of the above" is usually a pretty safe bet.

Use the True-False Technique to Change Perspective

What is the idea behind the true-false technique?

If you know your material but have a mental block about the multiple-choice format, you can gain a new perspective on a difficult question by using the true-false technique. Almost any multiple-choice question can be thought of as a series of true-false statements. Here's an example:

Before becoming president in 1857, James Buchanan was
a. married and divorced.
b. secretary of defense.
c. prime minister of Canada.
d. secretary of state.

This question and its options can be thought of as four true-false statements:

T (F) Before becoming president in 1857, James Buchanan was married and divorced.
T (F) Before becoming president in 1857, James Buchanan was secretary of defense.
T (F) Before becoming president in 1857, James Buchanan was prime minister of Canada.
(T) F Before becoming president in 1857, James Buchanan was secretary of state.

Viewing the question this way can sometimes make it easier to spot the correct answer. The true statement you find in the true-false statements you create usually contains the correct multiple-choice option.

Discard Foolish Options

What are distracters?

Some multiple-choice options are *distracters.* Whatever the reason for their inclusion, these foolish options are almost always good news for students. Exactly what the foolish option says is irrelevant. The important point is that you can eliminate it right away and pick the correct answer from the options that remain. Look at this example:

According to British tradition, the queen of England is not permitted to enter
a. West London.
b. Westminster Abbey.
c. the House of Commons.
d. the Indianapolis 500.

Option (d) is so silly that you can immediately cross it out. (Option [c] is correct.)

Choose the Middle Number from a Range of Numbers

Why is a middle number more likely to be the correct option?

Questions that use numbers as choices can be easily answered if you've memorized the correct number. But if you haven't really mastered your material or if you have a tough time with numbers in general, this kind of question can be a nightmare. If you have no other information to go on, you can increase your chances of guessing correctly by eliminating the highest and lowest numbers. Test writers usually include at least one number lower than the correct answer and one number higher. Using this "rule," you can eliminate half the options in a four-option question. For instance:

A water polo team has _____ players.
a. three
b. ten
c. seven
d. five

Even if you know nothing about water polo, you can use the midrange rule to eliminate two options and improve your odds from one out of four to one out of two. (The correct option in this case is [c].)

Finding the Best Match

What is the basic idea of a matching question?

Items in a matching test are usually divided into two columns and arranged in random order. Using a relationship that is normally explained in the directions, you systematically match the items in one column with the items in the other. (Consider the test shown on the following page.)

How are matching questions similar to multiple-choice questions?

Matching tests work like multiple-choice questions with an added dimension. You're faced with a *multiple* multiple choice. Instead of one stem and several options, you have several stems and several options.

What sort of complications does this extra dimension add?

This extra dimension adds extra complications as well. Matching carelessly or guessing prematurely can sometimes lead to a chain reaction of mistakes. If you make an incorrect match, you will deprive another item of its rightful match. This can aggravate your error by increasing the chances of another bad connection, which in turn can lead to another wrong match. Avoid this potential pitfall by making your matches carefully and by pairing up the items you are sure of before you begin guessing on items you're uncertain about. Answering matching questions effectively is mainly a matter of staying organized and saving time. A few strategies can increase your efficiency and reduce your confusion.

Groundbreaking Women of Science

Directions: Match the name of the scientist in the left-hand column with her discovery or distinction in the right-hand column by writing the proper letter in the space provided alongside each scientist's name. Use each item from the right-hand column only once.

	Scientist		Discovery or Distinction
_____	1. Rachel Carson	a.	Created the first programming language
_____	2. Mary Leakey	b.	Devised the system for classifying stars by brightness
_____	3. Marie Curie	c.	Garnered a Nobel Prize for her groundbreaking work in genetics
_____	4. Margaret Mead	d.	Earned fame for children's books but gained respect as a mycologist
_____	5. Jane Goodall	e.	First female physics instructor at Princeton University
_____	6. Grace Hopper	f.	Invented the pie chart, although she is best known as a nurse
_____	7. Georgia Dwelle Rooks	g.	First woman elected to the National Academy of Sciences
_____	8. Caroline Herschel	h.	First American woman in space
_____	9. Florence Sabin	i.	Founded the first hospital for African American women
_____	10. Florence Nightingale	j.	Invented the compiler, a critical element in most computer programs
_____	11. Chien-Shiung Wu	k.	Studied chimpanzees as a means of understanding human evolution
_____	12. Beatrix Potter	l.	Popularized anthropology with her book *Coming of Age in Samoa*
_____	13. Barbara McClintock	m.	Received two Nobel prizes for her work on radioactivity
_____	14. Annie Jump Cannon	n.	Uncovered fossils that helped unlock the origins of humankind.
_____	15. Sally Ride	o.	Founded the contemporary environmental movement
_____	16. Ada Byron King	p.	Teamed up with her brother to discover the planet Uranus

Answers: 1. o 2. n 3. m 4. l 5. k 6. j 7. i 8. p
9. g 10. f 11. e 12. d 13. c 14. b 15. h 16. a

Mark Off Matches to Avoid Redundancy

How does marking used items help?

This idea is so simple that it's often overlooked: Each time you match two items in a matching test, cross them off or mark them with a circle or an X. That way, when you move on to the next match, you'll have fewer items to read, and you won't be confused about which items you've chosen and which ones you haven't.

Match Shorter Items to Longer Ones

What's the benefit of matching shorter items to longer ones?

In most matching tests, the items in one column are longer than the items in the other. For example, a typical matching test might contain a column of terms and a column of definitions. In cases like these, you can save yourself some time if you set out in search of matches for the longer items instead of the reverse. In other words, the column you keep reading and rereading contains the shorter items. That way you need to read each long item only once. It's a case of the dog wagging the tail instead of the other way around.

Completing a Sentence

How are sentence-completion questions like multiple-choice questions?

Sentence-completion questions work like multiple-choice questions without the choice. A typical sentence-completion question consists of a partial sentence and one or more blanks. Your job is to read the sentences and to use both context and recall to determine what words belong in the blanks.

How do sentence-completion questions differ from multiple-choice questions?

Unlike multiple-choice questions, sentence-completion questions can't actually be considered objective questions, but because the sentence is incomplete and because the answer is seldom vague or ambiguous, most sentence-completion questions can be answered following the same basic procedure you use for answering bona fide objective questions. For example:

> Sentence-completion questions work like multiple-choice questions without the
> ____________________.

Why are there no real tricks for answering a sentence-completion question?

Because with a sentence-completion question the answer isn't there for you to choose, there are no real tricks to help you pick out the correct answer. But there are methods that enable you to clearly define the specific context of the question. When you do this, you zero in on the answer that will fill in the blank.

Clarify Ambiguity with a Specific Question

What is the best strategy for coping with ambiguous test questions?

Sometimes a question seems to have two or more reasonable answers. In these cases, you may need to clarify the kind of answer the question is seeking. The best strategy for coping with ambiguous questions is to raise your hand and ask a well-formulated, unambiguous question of your own to clear up the confusion. Consider this item:

> In 1901, at the age of forty-two, Republican Theodore Roosevelt became the country's ____________________ president.

In this example, both "youngest" and "twenty-sixth" would be reasonable answers, but it's unlikely the instructor would be looking for both. If you raised your hand and said, "I don't understand this question," you would probably get a response like "Do your best." But a well-thought-out, more specific question would probably be rewarded with a more helpful response. For example, if you asked, "Are you looking for a number?" the instructor's response would enable you to decide which of your two answers is expected.

Disregard the Length of the Blank

What effect does the length of the blank have on the desired answer?

Sometimes the length of the empty line equals the length of the answer expected. But in general there's no connection between the two. Pay attention to the words that are present, rather than to those that are missing, to come up with your answer for a sentence-completion question. Don't let the blank line distract you.

Treat Some Sentences as Two Separate Questions

What effect should two blanks have on your question-answering strategy?

Even students who aren't influenced by the size of one blank when answering a sentence-completion question may become flustered by a question that has two blanks. If the blanks are side by side, the question may be calling for a person's name or a place name. Paying attention to the question's context should help you confirm whether this is the case. But if the blanks in a sentence-completion question are widely separated instead of side by side, a different strategy is called for.

How do you treat two blanks that are widely separated?

The best way to treat two blanks that are widely separated is as if each occurred in a separate sentence. There may or may not be a direct relationship between the missing words, so make sure that each filled-in word makes sense in its own part of the statement. Here's an example:

> Although corn is second only to ____________________ as the most widely grown crop in the world, no one in Europe had even heard about corn until ____________________ returned from the New World.

In the first portion of the sentence, the word *corn* indicates that you're dealing with a grain. If you had read your textbook carefully (or if you used your common sense), you'd know the answer is wheat. The second blank demands a person's name: Columbus.

Moving Systematically Through the Test

How can you move systematically through an objective test?

Good students—those who understand what different objective questions require and have employed the most effective study methods—may still run into trouble unless they apply the same reasoned, organized approach they used in preparation to the process of taking the test. The only way to put what you know to good use is to move through the test systematically by reading carefully before you answer, by marking methodically, by paying attention to grammar, and by choosing the best answer instead of simply the first one that springs to mind. If you approach the test in this orderly fashion, you have an excellent chance of making the most of what you've learned.

Reading Before You Answer

What causes some students to skip over the obvious step of reading?

It may sound silly to suggest reading through the test before you begin answering. After all, the advice seems obvious. But because of anxiety and time pressure, common sense is often a casualty, and many students skip the directions and even rush through the questions without paying attention to what they say. In the long run, cutting corners on these crucial elements of a test could cost you time and points. Take a few moments to read the directions before you make a single mark on your test and to read each question carefully as well.

Read the Directions First

Why should you read the directions first?

It takes just a minute or two to read a test's directions, yet the little time you invest can often make a drastic difference in your score. Carelessness may do as much to torpedo a test as genuine ignorance. If, for example, the directions for a multiple-choice test say, "Mark the two best answers," but you pass over the directions and mark only one option, most of your efforts will have been in vain. Although simply reading the directions won't guarantee that you will pass, it will protect you from needless mistakes.

Read Questions Carefully

Why is it important to read each test question carefully?

Objective questions, no matter the type, are usually filled with information. Each word in the question is likely to be far more important than a word in an ordinary sentence. Nuances in phrasing, such as qualifiers and negatives, can have a huge impact on a question's meaning. For that reason, you must read each question carefully and thoroughly to pick up important details and the complete context.

How do qualifiers complicate a test question?

Cope with qualifiers. The English language has more than a dozen common qualifiers—including *always, most, equal, good,* and *bad.* We use such words regularly in writing and conversation, and test makers often deliberately insert them into objective test questions, especially true-false and multiple-choice questions. Qualifiers do precisely what their name implies: They complicate a simple statement or option by adding a qualification. The following two statements

> It *often* rains in Seattle, Washington.
> It *always* rains in Seattle, Washington.

are nearly identical. Yet the first one is true, while the second one is false. In this case, the only thing that differentiates the two statements is their qualifiers: *often* and *always.* If you read through these statements too quickly, you may overlook their qualifiers.

Now look at a multiple-choice example:

The head of a kettledrum is

a. struck only with wooden mallets.
b. always made of sheepskin.
c. often made of calfskin.
d. tightened once a day.

In this example, option (d) is just plain incorrect. Qualifiers indicate which of the three remaining options is correct. Without the qualifiers, all three options would be correct: Kettledrum heads *are* struck with wooden mallets; they *are* made of sheepskin; and they *are* made of calfskin. But because the qualifiers *only* and *always* overstate the case, options (a) and (b) are incorrect; while *often,* the qualifier in option (c), takes a more moderate stance and is therefore correct.

What are 100 percent words?

The qualifiers *only* and *always* in the first two options are good examples of 100 percent words. They imply that the statements are true 100 percent of the time. Such qualifiers almost always make a statement false; very few things in this world are 100 percent one way or the other. Although it is wise to watch out for these words, don't automatically consider a statement wrong because it contains one of them. To keep you honest and alert, some instructors occasionally use 100 percent words in true statements:

All stars are surrounded by space.
All human beings need food to survive.
No human being can live without air.

What is a good strategy for keeping tabs on qualifiers?

A simple and effective strategy for coping with qualifiers is to keep careful track of them by circling each one that appears in a test question. Circling the qualifiers helps ensure that you don't ignore them.

What is the Goldilocks Technique?

With the qualifiers circled, you can mentally substitute other words that will change the meaning of the question. This method is sometimes referred to as the Goldilocks Technique because you try out several qualifiers until you find the one that's "just right." Most qualifiers are clustered in groups, or "families." If you can find another family member that does a better job of completing the sentence, the original question is probably false or, in a multiple-choice question, is probably an incorrect option. The qualifiers in the families that follow may overstate a true-false statement, understate it, or make it just right. Memorize the six families. They will help you answer true-false questions and make the right choice among multiple-choice options.

- All—most—some—none (no)
- Always—usually—sometimes—never
- Great—much—little—no
- More—equal—less
- Good—bad
- Is—is not

What should you do when you recognize a word from a qualifier family?

Whenever one qualifier from a set is used in a true-false statement or a multiple-choice option, substitute each of the others for it in turn. In this way, determine which of the qualifiers from the family fits best (makes the statement just right). If that is the given qualifier, the answer is true; otherwise, the answer is false.

For example, suppose you are given this question:

T F All birds can fly.

Substituting the other qualifiers in the *all* family gives you these four statements:

Original Statement	**Related Statements**
All birds can fly.	Most birds can fly.
	Some birds can fly.
	No birds can fly.

The statement that begins with *most* is just right, but that is not the statement you were originally given. Therefore, the original answer is false.

What words are considered negatives in test questions?

Notice negatives. Negatives can be words such as *no, not, none,* and *never* or prefixes such as *il-,* as in illogical; *un-,* as in uninterested; and *im-,* as in impatient. Negatives are common in everyday speech and writing and almost as common in objective tests.

Why are negatives a problem?

Negatives cause problems in objective questions because, like qualifiers, they can easily be overlooked. This is particularly the case with negative prefixes, which have a way of blending in with the words they modify. For example:

> Because it is liquid at room temperature, mercury is indistinguishable from other metals.

If you read this sentence quickly, you may miss the letters *i-n* and mark the statement true. But if you read the statement carefully, you will realize that just the opposite is true.

What about sentences with more than one negative?

Objective questions that contain two or more negatives can be even more troublesome. Each additional negative shifts the meaning of the sentence in the opposite direction. For example, you would probably be able to mark this statement true without much difficulty:

> It is logical to assume that Thomas Edison's fame was due to his many practical inventions.

Yet you might have trouble with the sentence.

> It is illogical to assume that Thomas Edison's fame was not due to his many practical inventions.

Even though it is also true.

What should you do when you encounter negatives in a test question?

When you find negatives in objective questions, circle them. Then disregard them for a moment and try to gain the meaning of the question that remains. Finally, reread the sentence with the negatives included. Each negative you add reverses the meaning of the question. With two negatives, for example, the question's meaning should be the same as it was with none.

Hold Out for the Best Response

How can you tell a good response from the best response?

Some objective questions supply more than one good response, but in most cases there is only one best response. It's difficult to tell a good response from a best response unless you have read through the question completely. If you grow impatient and mark the first answer that sounds right, you risk missing the best answer.

Here's a multiple-choice example:

You would expect to find an aglet
a. on your foot.
b. in a nest.
c. in a small farming community.
d. at the tip of a shoelace.

An aglet is the cap, often made of plastic, at the end of a shoelace. If you read only part of this question, you might be tempted to pick option (a). That's a good choice, but if you read the whole question, you can easily see that it's not the best choice. Only with the question's entire context can you tell which option is a good answer and which is even better. (The best option is (d).)

Pay Attention to Grammatical Clues

How can grammar make test taking easier?

Although formats vary, all questions should follow the rules of grammar. This fact can help you narrow your choices by eliminating those possible answers that don't produce grammatically correct sentences. The only way to determine whether the rules of grammar are being followed or broken is by reading the entire question so you are able to get its total context. Consider this question:

The people of Iceland

a. a country located just outside the Arctic Circle.

b. are the world's most avid readers.

c. claim to be descendants of the Aztecs.

d. the capital, Reykjavik, where arms talks have been held.

If you race through this example, you might be tempted to mark either (a) or (d) as the correct response. Indeed, Iceland is a country located just outside the Arctic Circle, and Reykjavik, the capital, has been the site of important arms negotiations. But if you take the time to read the entire question, you can see that these two responses do not complete the stem grammatically. Response (a) is missing a verb, and response (d) is missing any grammatical connection to "The people of Iceland." That leaves (b) and (c) as the only legitimate options. (The correct answer is [b].)

Grammatical clues are even more helpful in sentence-completion questions, where your response must be recalled instead of chosen from a list of possible answers. For example, you don't have to be an expert in astronomy to complete this sentence:

> Although about 75 million meteors enter our atmosphere each day, on the average only ____________________ of them ever reaches the ground.

Because *reaches* is a singular verb, the only correct answer is *one.* (Otherwise the question would have read *reach,* the plural form.)

Marking Methodically

How do you move methodically through an objective test?

As with many things, a key to successfully answering objective questions is to move through the test methodically. Don't tackle the questions in a slapdash fashion. Instead, cycle through each section from start to finish, marking only the "sure things" at first and making an intelligent guess if you're still stymied by a question the third time around.

Mark Only the "Sure Things" at First

Why should you mark only the "sure things" at first?

On your initial pass through the test, mark only those answers you are sure of. This is especially crucial in matching questions, where one mistake can set off a chain reaction of incorrect answers. If a question has you stumped at first, don't feel compelled to answer it right away. And don't pick a temporary answer with the thought that you can come back and change it later. You may not have time, and even if you do, you may not be able to distinguish your uncertain answers from your certain ones.

What should you do if the correct answer doesn't immediately come to mind?

If an answer doesn't come to mind right away, circle any qualifiers or negatives, eliminate any choices you know are incorrect, and then move on to the next questions. These markings will provide you with a head start on your second pass.

Guess the Third Time Around

Is there any benefit to guessing?

Except when there's an extra penalty for incorrect answers, guessing is always better than simply leaving a question blank. A question unanswered guarantees a zero, whereas a guess may score some points. Furthermore, if you know something about the material and have given it some genuine thought, you should be able to make an intelligent guess. Intelligent guesses are always superior to random ones. Consider this particular sentence-completion item:

> You can travel by ship from New England to Florida without ever entering the usually rough open seas by using a system of rivers and canals called the ____________________.

If you don't know the official name for the system, this sentence is long enough and descriptive enough to help you come up with a good guess. You might call it the "Inland Waterway." That's not the exact name, but it is very close, and you would likely receive partial credit for it. (The answer is Intracoastal Waterway.)

FINAL WORDS

How are objective tests like pasta?

If you're like many students, you have a favorite type of objective question as well as a type that you dread. To some extent, that's only natural. But as you prepare for your next objective test, keep this in mind. Objective tests are really a lot like pasta. Some people love rigatoni, some like capellini, and others would rather stick with plain old macaroni. Yet despite their differences, most pastas share the same basic ingredients. It's only the shapes that separate one from the other. If you have the dough to make macaroni, you can always make rigatoni or capellini instead. In the same way, if you feel confident and well prepared for a true-false test, you're probably in a better position to take a matching or multiple-choice test than you realize. Learn to appreciate the shared ingredients of all objective questions, and there's a good chance that your next test will be a little less traumatic.

HAVE YOU MISSED SOMETHING?

SENTENCE COMPLETION

Complete the following sentences with one of the three words listed below each sentence.

1. Negatives and qualifiers should be ________________.

   avoided **circled** **defined**

2. In matching questions, the fewer the remaining choices, the better your chances of being ________________.

   incorrect **correct** **alert**

3. The greatest threat of negatives is that they can be easily ________________.

   replaced **overlooked** **dismissed**

MATCHING

In each blank space in the left column, write the letter preceding the phrase in the right column that matches the left item best.

_____ 1.	Negative	a.	Statement that starts off a multiple-choice question
_____ 2.	Guessing	b.	One of the choices that make up a multiple-choice question
_____ 3.	Context	c.	Mastering possible test material from both sides
_____ 4.	Stem	d.	Is almost always better than leaving a question blank
_____ 5.	"Sure thing"	e.	Should be marked on the first pass through the test
_____ 6.	"Switch hitting"	f.	Best choice when stumped by a true-false question
_____ 7.	True	g.	Usually reverses the meaning of a true-false statement
_____ 8.	Option	h.	Is provided when you read the entire question

TRUE-FALSE

Write T *beside the* true *statements and* F *beside the* false *statements.*

_____ 1. For a statement to be marked true, it must be entirely true.

_____ 2. Some multiple-choice questions ask you to pick more than one answer.

_____ 3. The stem of a multiple-choice question is always an incomplete statement.

_____ 4. True-false tests generally contain more true statements than false ones.

_____ 5. The length of the blank dictates the size of the answer in a sentence-completion question.

MULTIPLE CHOICE

Choose the word or phrase that completes each sentence most accurately, and circle the letter that precedes it.

1. A multiple-choice question can be viewed as a series of
   a. stems.
   b. qualifiers.
   c. true-false statements.
   d. decoys or distracters.

2. One way to think of a matching question is as
   a. a multiple-choice question without the choice.
   b. a true-false question in two dimensions.
   c. a multiple multiple-choice question.
   d. none of the above.

3. Reading the entire question should help you
   a. detect grammatical clues.
   b. take advantage of context.
   c. select the best response.
   d. all of the above.

4. If at first a question has you stumped, you should
   a. move on to another question.
   b. ask a clarifying question.
   c. pick a temporary answer.
   d. cross out any negatives or qualifiers.

5. A sentence-completion question with two widely separated blanks should be treated as
   a. a true-false question.
   b. two questions.
   c. a decoy or distracter.
   d. an essay question.

SHORT ANSWER

Supply a brief answer for each of the following items.

1. Compare and contrast the four basic types of objective questions.
2. Explain the Goldilocks Technique.
3. What is the chain reaction associated with matching tests?
4. What is the purpose of the true-false technique?

WORDS IN CONTEXT

From the three choices beside each numbered item, select the one that most nearly expresses the meaning of the italicized word in the quote. Make a light check mark (✓) next to your choice.

Nations are not ruined by one act of *violence,* but gradually and in an almost *imperceptible* manner by the *depreciation* of their circulating currency, through its *excessive* quantity.

—Nicolaus Copernicus (1473–1543), Polish astronomer

1. act of *violence*	damaging force	enormous effect	gigantic power
2. almost *imperceptible*	unknown	controlled	unseen
3. the *depreciation*	stable value	decreased value	increased value
4. *excessive* quantity	commercial	inordinate	required

Another advantage of being rich is that all your faults are called *eccentricities.*

—Anonymous

5. called *eccentricities*	creative	peculiarities	inventive

No man can be *conservative* until he has something to lose.
—James P. Warburg (1896–1969), American publicist

6. *conservative* objective cautious judicious

When a subject becomes totally *obsolete* we make it a required course.
—Peter Drucker (1909–), American business philosopher and author

7. totally *obsolete* indispensable outdated necessary

THE WORD HISTORY SYSTEM

Enthrall
literally, to enslave

enthrall en-thrall' *v.* 1. To hold spellbound; captivate. 2. To enslave.

Enthrall presents another case of a word the original and literal sense of which is cruel, but the modern, figurative use of which is much more pleasant. When we say that we are *enthralled* by a song, or a book, or something else with captivating charm, it is interesting to remember that the original meaning of the word was "to enslave." *Thrall* is Anglo-Saxon for "slave." To *enthrall* meant, therefore, "to enslave," "to reduce to the condition of a thrall." The literal sense of "enslave," "make captive," easily yields a figurative sense, "captivate the senses," "hold spellbound," "charm," as with a song or a story.

Reprinted by permission. From *Picturesque Word Origins* © 1933 by G. & C. Merriam Co. (now Merriam-Webster, Incorporated).

Appendix: Answers

Chapter 1 Controlling Your Time

HAVE YOU MISSED SOMETHING?
Sentence completion: 1. game plan 2. daytime 3. term
Matching: l. g 2. f 3. d 4. b 5. h 6. c 7. a 8. e
True-false: l. F 2. T 3. F 4. T 5. T
Multiple-choice: l. b 2. b 3. c 4. c 5. d

WORDS IN CONTEXT
1. strength 2. sorrowfully 3. death 4. worth 5. inexplicable 6. associated 7. included

Chapter 2 Adopting a Note-Taking Mindset

HAVE YOU MISSED SOMETHING?
Sentence completion: 1. context 2. ideas 3. anticipation
Matching: l. c 2. e 3. b 4. g 5. d 6. a 7. h 8. f
True-false: l. T 2. F 3. T 4. F 5. T
Multiple-choice: 1. d 2. d 3. a 4. c 5. d

WORDS IN CONTEXT
1. dictionary 2. unbeatable 3. stern

Chapter 3 Taking Effective Notes

HAVE YOU MISSED SOMETHING?
Sentence completion: 1. forgetting 2. signposts 3. questions
Matching: l. b 2. f 3. d 4. h 5. g 6. e 7. c 8. a
True-false: l. F 2. T 3. F 4. T 5. T
Multiple-choice: l. b 2. c 3. b 4. c 5.d

WORDS IN CONTEXT
1. refinement 2. relieve 3. reasonable 4. dead end

Chapter 4 Answering Objective Tests

HAVE YOU MISSED SOMETHING?

Sentence completion: 1. circled 2. correct 3. overlooked
Matching: 1. g 2. d 3. h 4. a 5. e 6. c 7. f 8. b
True-false: 1. T 2. T 3. F 4. T 5. F
Multiple-choice: 1. c 2. c 3. d 4. a 5. b

WORDS IN CONTEXT

1. damaging force 2. unseen 3. decreased value 4. inordinate
5. peculiarities 6. cautious 7. outdated